Starring
T. rex!

Dinosaur Mythology and Popular Culture

José Luis Sanz

Starring
T. rex!

TRANSLATED BY PHILIP MASON

INDIANA
University Press
Bloomington & Indianapolis

This book is a publication of
Indiana University Press
601 North Morton Street
Bloomington, Indiana 47404-3797 USA

http://iupress.indiana.edu

Telephone orders	800-842-6796
Fax orders	812-855-7931
Orders by email	iuporder@indiana.edu

*The paper used in this publication meets the minimum
requirements of American National Standard for Information
Sciences—Permanence of Paper for Printed Library
Materials, ANSI Z39.48-1984.*

MANUFACTURED IN THE UNITED STATES OF AMERICA

Library of Congress Cataloging-in-Publication Data

Sanz, J. L. (José Luis)
  Starring T. rex! : dinosaur mythology and popular culture /
José Luis Sanz ; translated by Philip Mason.
     p. cm.
Includes index.
  ISBN 0-253-34153-1 (cloth : alk. paper) —
ISBN 0-253-21550-1 (pbk. : alk. paper)
  1. Dinosaurs in mass media.   2. Dinosaurs—Mythology.
I. Title.
  P96.M6 S264 2002
  567.9—dc21
                              2002004559

1   2   3   4   5   07   06   05   04   03   02

For Nuria, sweet companion for traveling to Jurassic and other paradises

For Mercedes, who shares with me the hope that dinosaurs will never leave our dreams

For Loli, Jorge, Javier, and Alejandra, for many years of affection, paleontology, mine engineering, and guacamole

CONTENTS

ACKNOWLEDGMENT

The publisher wishes to thank M. K. Brett-Surman for his assistance in the preparation of this edition.

INTRODUCTION

Dinosaurs make up a singular group of terrestrial vertebrates that lived in our remote past. Most became extinct some 65 million years ago. However, birds—the dinosaurs' descendants—have managed to survive up to our day. All kinds of forms disappeared: bipeds and quadrupeds, carnivores and herbivores, with sizes ranging from that of a turkey to 10 times as large as an elephant.

In recent years, the number of paleontologists who study dinosaurs has increased significantly. The amount and importance of the information about dinosaurs' evolutionary history that is currently available was absolutely unthinkable just a few decades ago. Like many other objects science focuses its attention on, dinosaurs may be analyzed from the sociocultural point of view, apart from strictly paleontologic study. The sociocultural reach of dinosaurs can be considered as the result of a process that moves information out of the scientific arena and into the popular consciousness. Ideas are propagated, maintained, and modified thanks to the mass media—newspapers, popular literature, comic books, television, and movies—and the result is clear: they have managed to push these Mesozoic vertebrates into mythologic realms. In this way, dinosaurs have become a symbol, and occasionally even the paradigm, of some very specific concepts.

Two of the clearest examples that illustrate this point are the interpretation of dinosaurs as something obsolete and out of date, and the supposed teachings and warnings they provide with respect to human destiny. Both symbolic views probably come from the biologic interpretation of dinosaurs in the 1940s and 1950s. In those years, the general opinion among paleontologists was that brontosaurs and tyrannosaurs were clumsy, heavy, stupid beasts. They were aberrations of nature, inevitably doomed to disappear, and nature thus took care of eliminating them. Since then, there has always been a symbolic association between dinosaurs and the extinction of living organisms. That association, taken with the crisis in the collective consciousness springing from fear of a nuclear holocaust, leads to a warning: humans could disappear from the planet, just like the dinosaurs! Dinosaurs have also been consistently used as a symbol of repressive ideas and backwardness, including in politics. Even so, the most widespread mythical dimension of the dinosaurs is their characterization as dragons.

The phenomenon of dinosaur mythology has developed thanks largely to the existence of many popular-science publications, which have increased their circulation enormously in recent years. In Western societies, especially the United States, some children are more familiar with brontosaurs, hadrosaurs, and tyrannosaurs than with lions and elephants. Whenever human beings make contact with the distant past in movies, comic books, or science fiction literature, dinosaurs are there.

Dinosaurs have an indisputable commercial value in our society, including their use in advertising and marketing, logos, and toys, as well as in a multitude of other manufactured products. An organization called the Dinosaur Society was founded in the United States at the beginning of the 1990s, with one of its goals being to attempt to create a bridge between professional dinosaur paleontologists and the business world that was commercializing dinosaurs' images. The organization, for instance, had a major role in shaping Steven Spielberg's movie *Jurassic Park*. Clearly, all these phenomena reflect the large impact and sociocultural importance achieved by dinosaurs. Their impact comes in characteristic waves of increas-

A pre-issue artist proof U.S. postage stamp from 1997 featuring *Daspletosaurus*. This was only one of three series of dinosaur postage stamps issued by the U.S. Postal Service in the 20th century. Artwork by James Gurney.

ing popularity and societal demand for dinosaurs, a phenomenon known as dinomania.

This book is divided into two main parts. The first is dedicated to the study of the historical process that has generated the current complex mythology of the dinosaurs; the second is concerned with analyzing the structure of the myth. The process that generates myth is analyzed within a series of historical stages beginning with the 17th century and finishing with the great wave of dinomania created by Spielberg's *Jurassic Park* and its sequels, *The Lost World: Jurassic Park II*, and *Jurassic Park III*. The second part (starting with chapter 9) lays out an analysis of the narrative discourses—in film, comics, and literature—that include dinosaurs. Among other themes, this part analyzes dinosaurs' appearance in fantasy tales; the relationships—usually difficult—between humans and dinosaurs; the morphology and way of life of dinosaurs presented in the popular media and the techniques used in movies to recreate those details; and the relationship among dinosaurs, dragons, and the famous creatures of Japanese cinema.

In sum, this work provides an analysis of the singular sociocultural phenomenon that has been generated by a group of animals from our remote past—creatures that ever since their discovery have deeply fascinated human beings. This book is meant for anyone who has ever felt

Part of the personal memorabilia collection of dinosaur author Donald F. Glut. Photograph by M. Brett-Surman.

Pentaceratops, a bronze sculpture by the late dinosaur artist Dave Thomas. This model can be seen outside the New Mexico Museum of Natural History in Albuquerque.

a special thrill while contemplating the skeleton of *Diplodocus* in a museum, or seeing Godzilla launch his deadly radioactive breath over Tokyo.

Starring
T. rex!

Surprise: The Origin of the Myth

The first written information about the discovery of dinosaur bones comes from the 17th century. In 1677, Robert Plot, the first curator of the Oxford Ashmolean Museum, published his *Natural History of Oxfordshire*. In this book, he described and illustrated a distal fragment of a large femur that Plot interpreted as belonging "to a man or, at least, to some other animal," perhaps, according to him, to one of the elephants brought to Great Britain by the Romans. This bony relic, whose current whereabouts are unknown, probably came from a carnivorous dinosaur (perhaps *Megalosaurus*). R. Brookes gave it the name *Scrotum humanum* in 1763. This binomial represents the first published name of a dinosaur, but would now be considered a *nomen oblitum* (forgotten name) because it has not been used by any author in more than 50 years.

In any case, these first dinosaur remains to have written and graphic proof had already become the object of ideological interpretation. In 1768, J. B. Robinet assessed these remains from a seminalist perspective: fossils could grow in the interior of the Earth from seeds of a living organism. Robinet described this *Scrotum humanum* as truly belonging to a man who had developed in the subsoil, and he provided anatomical details of the man's supposed testicles.

From the end of the 17th century to the beginning of the 19th, many discoveries of dinosaur remains were made in western Europe (mostly England and France) and North America. Nevertheless, the first modern descriptions of dinosaurs come from the beginning of the 19th century. We understand as *modern* those interpretations of dinosaur remains that for the first time paid attention to the unusual nature of the finding. This basically means that dinosaur bones were interpreted as belonging to a probably extinct group of enormous reptiles of dimensions unknown today in terrestrial vertebrates.

Two English naturalists were responsible for these pioneering studies. In 1824, the University of Oxford don Reverend William Buckland (1784–1856) published a report on the remains of the large carnivore that he named _Megalosaurus._ In 1825, the doctor Gideon Algernon Mantell (1790–1852) proposed the name of _Iguanodon_ for the huge animal whose remains were found in the Mesozoic strata of Tilgate Forest (Sussex). These same sites yielded bones of another enormous reptile in 1832, described by Mantell in 1833 and given the name _Hylaeosaurus._

The discovery of these first three dinosaurs (_Megalosaurus, Iguanodon,_ and _Hylaeosaurus_) gave rise to a new idea. It seems fairly clear that in the 1830s, the belief began to become established that before humans existed, in Jurassic and Cretaceous times, the Earth was inhabited by large, and previously unknown and unsuspected, terrestrial reptiles. The naturalist who immediately after Buckland and Mantell provided more information about these new and enormous fossil beasts was Richard Owen (1804–1892), the first director of the British Museum and a personal friend of Queen Victoria. Owen was not only responsible for moving scientific investigation of dinosaurs forward, but was also the author of the first general interpretation of the group (later used to back up an ideology, as we shall see). Also, to a large extent, he was the driving force behind a process of popularization of dinosaurs, which gave rise to definite opinions in the popular consciousness of his time.

In 1842, during a meeting of the British Association for the Advancement of Science, Owen proposed the term "Dinosauria." The Victorian naturalist conceived of the dinosaurs as a special group of fossil reptiles with some advanced characteristics that in some ways were similar to those of mammals. Owen reconstructed _Megalosaurus_ and _Iguanodon_ as large quadrupeds resembling a rhinoceros with its extremities made vertical. This was a different model from those previously put forward, which conceived of dinosaurs as being enormous lizards.

The reasons for this new interpretation are obvious. Owen, one of the best anatomists of his age and a strict follower of the methodology of Georges Cuvier, verified that the new group of large reptiles had characteristics that were absent from present-day forms. Lizards and crocodiles have two sacral vertebrae, whereas dinosaurs have five. Also, the morphology of the ribs implies deep, tall thoracic cavities and an

To celebrate the New Year, a dinner was served on 31 December 1853 inside a life-size *Iguanodon* made of metal and cement by B. Waterhouse Hawkins under the direction of Richard Owen. Owen's place, raising a toast inside the head of the enormous beast, is clearly allegorical, referring to his status as leader in the study of dinosaurs. On this occasion a song was written whose chorus went: "The jolly old beast / Is not deceased / There's life in him again." From The Illustrated London News, 1854.

animal that bore itself well off the ground on long, powerful limbs. Everything seemed to indicate that dinosaurs were terrestrial beasts of large dimensions; the closest references were the large phytophagous (plant-eating) mammals (such as elephant and rhinocerous). In this way, the Owenian model of dinosaurs was based on direct deduction from observations of comparative anatomy. For this reason, the view of A. Desmond, an historian of paleontology, which argued that Owen needed a race of super-reptiles and therefore created it, is doubtful.

According to Desmond, Owen invented his model of dinosaurs in order to attack the transformist ideas of Lamarckism. Lamarck considered that there was an increase in the structural complexity of living creatures, from the simplest to the most highly organized. Owen rejected this proposal, supporting his ideas on the basis of the fact that the most complex (highest) forms of reptiles were not the present-day ones but rather the huge beasts of the Mesozoic, the dinosaurs. Clearly, we are faced here with an ideological assessment of dinosaurs. However, it is likely that the Owenian model was not invented for such purposes, but rather that it was built up from strict empirical observations and was subsequently used as evidence to corroborate an ideological position.

Research into dinosaurs carried out by naturalists such as Buckland, Mantell, and above all Owen provoked great expectations in the public opinion of the time, particularly in Britain and to a lesser extent in France and other countries of continental Europe, as well as North America. The considerable sociocultural impact of dinosaurs in Victorian Britain was due to the collaboration between Owen and the artist Benjamin Waterhouse Hawkins (1807–1889), who drew and sculpted the first three-dimensional reconstructions of dinosaurs under Owen's direction. Today, in a small park in the London suburb of Sydenham,

The Waterhouse Hawkins/Owens reconstruction of *Hylaeosaurus*, which was conceived like a large terrestrial mammal, similar to a rhinoceros. Displayed in 1853 at Crystal Palace, this restoration can still be seen in Hyde Park, London.

Current reconstruction of *Iguanodon,* a bipedal animal that could also move quadrupedally. Model created by Javier Pérez Valdenebro.

you can still see his models of *Megalosaurus* and *Iguanodon* that appeared at the Crystal Palace exhibition (1853). Dinosaurs, as well as other animals from our distant past, sculpted by Hawkins, were presented as "antediluvian monsters." This term, the use of which is still occasionally seen in today's media, implies the belief in a devastating flood of supernatural origin that managed to eliminate the dinosaurs and other beasts of long ago.

During the second third of the 19th century, dinosaurs were conceived of as enormous reptiles with appearances similar to large terrestrial mammals. Dinosaurs were contemporaries of the "dragons of the air" (pterosaurs) and the marine "snakes" and "monsters" (plesiosaurs and ichthyosaurs). The public imagination was captured by a sense of surprise at the discovery of a group of absolutely unimaginable fossil beasts. In the next chapter, we will see how this initial feeling of astonishment gave way, during the final third of the 19th century and the beginning of the 20th, to one of certainty about their existence and of the "everydayness" of dinosaurs—that is, their regular presence in society.

2

Dinosaur Hunters: They Really Existed!

The stage for the second phase of the genesis of dinosaur mythology shifted from Europe to the United States, a country that has clearly been most responsible for the extent to which the myth has developed. The tendency for U.S. culture to export itself and the formidable potential of the country's social media has ensured that the myth has become installed in many other countries, especially those with close economic and sociocultural affinities to the United States.

The first dinosaurs to be discovered in the United States were found in the upper Cretaceous strata of Gloucester County (New Jersey) in 1787. Around 1802, dinosaur footprints were found in the Triassic (nowadays dated from the Lower Jurassic) levels. As a result of their tridactyl (three-fingered) morphology, they were thought to be the prints of birds, probably from the crow that Noah sent out from his ark or produced by the feet of poultry.

It is clear that the footprints (tracks) of dinosaurs have historically given rise to nonscientific interpretations that may be identified in a wide range of beliefs in European and American folklore. From this point of view, this type of indirect fossil evidence has three main characteristics: (1) The alignment of footprints (trackways) are easily interpreted as being the result of an animal's walk. (2) The majority are three-toed footprints, sometimes quite large in size; footprints of 30 to 50 cm in length abound in the ichnologic record. (3) The impressions always occur in solid rock, and not in mud or soft sediment. These three characteristics allowed an initial approach to the analysis of the generation of the myth of dinosaur footprints. The significance of the first point is obvious: they provide the observer with a direct interpretation of the phenomenon. The second and third explain why the animal could not be found in the real world and why an explanation required recourse to be made to supernatural powers (no *real* animal could leave such footprints in hard rock).

Various places in western North America where dinosaur tracks are to be found also feature Native American paintings. The U.S. dinosaurologist Jim Farlow suggested that the famous dinosaur footprint sites of the Paluxy River (Lower Cretaceous, Texas) could have been associated with the "Thunderbird" by the Native Americans. The North American Hopis use ritual skirts for their snake dance that are decorated with dinosaur tracks. In 1941, Kirchner suggested that the tracks of *Cheirotherium*—produced by a primitive archosaurian reptile with a crocodile-like appearance—from Siegfriedsburg (German Triassic) may have been associated with the legend of Siegfried and the dragon (see chapter 24).

In Portugal, one of the best-known areas for dinosaur tracks is the Lower Cretaceous of the Bay of Lagosteiros (Cape Espichel). The devout tradition of the place led to the organism responsible being identified as the mule that transported the Virgin Mary inland from the coast. In Spain, the country people and shepherds of the region of La Rioja Baja (e.g., Enciso, Navalsaz, Préjano) attributed the numerous ichnites of the area to the hooves of Saint James's horse. The traditional interpretation of the tridactyl footprints as being produced by the hooves of an equine species in Spain and Portugal may seem surprising, given the great morphologic diversity that exists. Nevertheless, we have to bear in mind the need to square a phenomenon inexplicable in terms of Nature with the supernatural tradition (in this case, a religious one). The pre-Christian people of La Rioja would doubtless have had different interpretations of the same phenomenon.

In the middle of the 19th century, the cleric Edward Hitchcock (1793–1864), professor of natural theology and geology at Amherst College, studied the Lower Jurassic footprints from the Connecticut Valley. In 1858, Hitchcock published his *Ichnology of New England: A Report on the Sandstone of the Connecticut Valley, Especially Its Fossil Footmarks*. Hitchcock proposed in this monograph that the Triassic tracks were generated by giant birds—ostriches, emus, and even forms similar to the extinct New Zealand moa. Nevertheless, certain traces with the impression of a tail led Hitchcock to suppose that some of these birds must have had particular reptilian characteristics. This interpretation shows considerable intuition concerning the real authors of the tracks: bipedal dinosaurs that were close relatives of the birds. The relationship between birds and dinosaurs was also proposed by

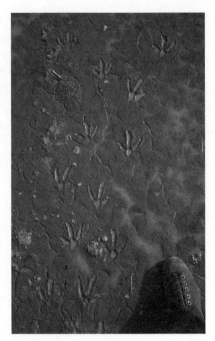

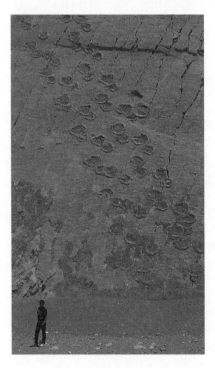

Ichnology is the science of the study of the tracks and footprints of living organisms. Paleoichnology is their historic equivalent—the study of ichnological evidence of animals of the past. The footprints of dinosaurs can be found throughout the world, although the region of Cameros (La Rioja, Burgos and Soria, Spain) is one of the most outstanding areas known. (*top left*) Tracks of large theropod dinosaurs at the spectacular site of Los Cayos (Cornago, La Rioja, Spain). (*top right*) Current footprints of a rhea in soft mud in Patagonia (Argentina). Their similarity to those of the great theropod dinosaurs is evident. It is therefore not surprising that E. Hitchcock interpreted the famous tracks from the Connecticut Valley as belonging to large flightless birds. (*left*) Tracks of enormous sauropods from the Upper Cretaceous of the El Molino Formation on the limestone slopes of Sucre (Bolivia). Courtesy of Enrique Díaz Martínez.

Dinosaur tooth fragments for which the U.S. dinosaurologist Joseph Leidy proposed the term *Trachodon*. These fossils are from the Upper Cretaceous of Montana. Nowadays, *Trachodon* is not considered to be a valid dinosaur name, as evidence from teeth is not sufficient for the individual identification of genera within a family. This rule is valid for most dinosaurs. Courtesy of the Philadelphia Academy of Sciences.

Thomas Henry Huxley (1825–1895), Charles Darwin's famous bulldog, and as we shall see later, this is a commonly accepted hypothesis in our times. Huxley's opinion of the footprints from Connecticut partially coincided with that of Hitchcock, who argued for the existence of a mixed community of large running birds and dinosaurs.

Joseph Leidy (1823–1891) is generally considered to be the initiator of dinosaur studies in the United States. In 1856, he described teeth from a dinosaur similar to *Iguanodon*, which he named *Trachodon*. In 1858, in Haddonfield (New Jersey), some relatively abundant bony remains of a type of dinosaur that Leidy named *Hadrosaurus* were unearthed. These remains led Leidy to propose that dinosaurs with this type of dentition (*Iguanodon, Trachodon, Hadrosaurus*) were probably animals similar to kangaroos in their posture and mode of movement — a radically different model from that of the heavy quadrupeds that Owen supposed. Huxley had certainly already reiterated the bipedal character of *Iguanodon*, but it seems that his proposals were not given consideration. The new evidence provided by the *Hadrosaurus* material from

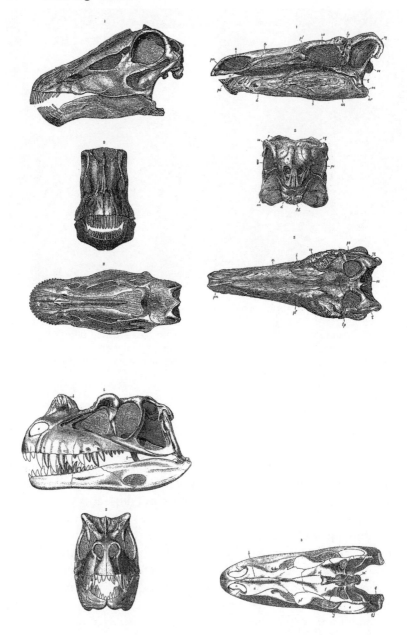

O. C. Marsh proposed many of the most popular genera of dinosaurs. Among these are the stegosaur *Stegosaurus* (above right), the sauropod *Diplodocus* (above left), and the basal theropod *Ceratosaurus*. Illustrations of the skulls of these three dinosaurs were published by Marsh in his book *The Dinosaurs of North America* (1896).

New Jersey confirmed the nonquadrupedal nature of the dinosaur named by Mantell. The discovery of complete skeletons of *Iguanodon* in the Belgian town of Bernissart in 1878 settled the issue in favor of bipedalism.

As we will see later, Leidy's critical stance in the face of the proposals of the English naturalists was also a constant feature adopted by two famous U.S. paleontologists, Edward Drinker Cope (1840–1897) and Othniel Charles Marsh (1831–1899). Both scientists embodied the essence of what has come to be known as the *dinosaur hunter*, best viewed in the context of the North American Wild West, where these two paleontologists collected a vast quantity of dinosaur and other vertebrate bones, even in areas where there were hostile Native Americans. This aura of adventure certainly contributed to the early the popularity of dinosaurs in the land of Coca-Cola. The romance of prospecting and excavating fossil vertebrate remains captured the collective North American imagination, much as the adventures of "Buffalo" Bill Cody had. This led to the dinosaur hunters being included within that greatest of North American national epics: the exploration and colonization of the territories of the West.

It is difficult to evaluate the significance of Cope's and Marsh's work, but two aspects stand out. First, the great quantity of fossil material obtained, frequently of complete skeletons, allowed access to an enormous amount of new information about dinosaurs that, in some cases, contradicted the interpretations of the European naturalists. In this respect, an unmistakable sense of superiority comes across in Marsh's (1895) opinions about the conceptual errors of the European paleontologists. Cope and Marsh undoubtedly comprised the vanguard of dinosaur research during the last quarter of the 19th century, and this phenomenon made its mark on the opinion of the North American public. Cope and Marsh were the leading edge at a critical moment in the history of dinosaur research that must be seen and evaluated in the context of a broader, interrelated phenomenon: the impact of paleontology in the United States.

The search for new forms of dinosaurs by paleontologists such as Cope and Marsh should be considered within the context of the prestigious role played by paleontology in the cultural development of the United States in the early decades of the 1900s. We should not forget that we are talking about a country whose third president, Thomas Jefferson, exhibited mastodon bones in the White House, and where a newspaper, the *New York Tribune*, printed the lectures of Harvard zoologist Louis Agassiz. It is no surprise that the famous U.S. paleobiologist

George Gaylord Simpson said that paleontology had yielded the greatest contribution to the history of American culture. In this context, Cope's and Marsh's numerous contributions to dinosaurology acquire a new dimension—one that has had an influence on the search for an authentic national identity and the recognition of the rest of the community of nations. The discoveries in the United States astonished the world.

All things considered, the work of Cope and Marsh, their position within an epic of the building of the nation, and their modern and innovative character explain the enormously broad spread of dinosaurs through culture toward the end of the 19th century and at the beginning of the 20th century. Their significance in the construction of the myth seems obvious: dinosaurs became established in the popular imagination of the United States as something commonplace and intimately related to the American way of life. They are organisms from the distant past that have a real presence in our own day and age. In short, U.S. society—and shortly after, other Western nations—accepted as something entirely normal that dinosaurs really existed and that they were not beasts with a lizard- or mammal-like appearance, as conceived of by Victorian naturalists. Dinosaurs make up a diversity of forms that represent an alternative world of terrestrial vertebrates. This alternative character was probably one of the dinosaurs' greatest initial popular attractions and encouraged their establishment in popular culture in the early decades of the 20th century.

3

Arthur Conan Doyle, Lost Worlds, and Cavemen

Up to this point, I have looked at the activities of paleontologists and the beginnings of a dinosaur cultural mythology. A decisive stage in this process has its roots in the first two decades of the 20th century and owes its development to science fiction cinema and literature. Naturally, its origin is centered on the avalanche of direct and speculative information generated by the activity of Marsh and Cope, and, to a lesser extent, by the discoveries in Europe.

The year 1912 saw the publication of a work of literature that would have a great influence on the subsequent development of the myth: *The Lost World* by Arthur Conan Doyle (1859–1930). The subject of the lost world—the persistence in the present day of the fauna and flora of the remote past—in this novel by the creator of Sherlock Holmes has some precedents. In 1833, the Swiss draftsman Rodolphe Toepffer (1799–1846) published his novel *Voyages et Aventures du Docteur Festus* (*Journeys and Adventures of Doctor Festus*). In this tale, the main character is launched into space by means of a windmill and goes into orbit around the Earth. During his journey, Dr. Festus observes that the poles are pierced by an enormous hole, at the bottom of which can be seen the magma of the interior of the planet. The heat has allowed prairies to flourish around the hole, with their perimeters bounded by ice. These prairies are inhabited by mammoths and mastodons.

Jules Verne's (1828–1905) *Voyage au Centre de la Terre* (*Journey to the Center of the Earth*) appeared in 1864. The novel tells of the adventures of Professor Lindenbrock, who, with his nephew and an Icelandic guide, reached the center of the Earth by descending the crater of the Sneffels Volcano, and appeared on the surface once again through Stromboli (Sicily). The members of the expedition, after many adventures, discover an interior sea, where they witness a battle between a plesiosaur and an ichthyosaur. The entire work, as is typical of Verne's stories, has a surprisingly didactic tone—in this case, being that of a mineralogy and paleontology manual. The description of the anatomy and behavior of the two large marine reptiles in the episode describing their fight is probably based on Louis Figuier's work, *La Terre Avant le Déluge* (*The Earth before the Flood*), which appeared in 1863 and provided a faithful interpretation of the paleontologic knowledge of the period. The expedition members even see the spout of air and water expelled by the huge ichthyosaur, a trait possibly inspired by an illustration in Figuier's novel.

The two precursory stories by Toepffer and Verne are clearly related to the novel of utopian journeys and situate the unknown location in physically improbable places—here, at the poles and inside the planet, respectively. Doyle's scenario situates the lost world in a real but little-explored place, the Brazilian forest. Significantly, this place is populated by an essentially European dinosaur (*Iguanodon*) and by one

The formidable fight between an ichthyosaur (on the left, ejecting two spouts of water) and a plesiosaur, from Louis Figuier's book *The Earth before the Flood* (1863). This illustration probably inspired Jules Verne's description of the struggle between two enormous marine non-dinosaur reptiles in his novel *Journey to the Center of the Earth* (1864). The drawing, like all the illustrations in the first edition of Verne's novel, is by the French artist Riou, who thus established a direct link between the popularization of the paleontology of the period and the fictional story.

of the North American forms that had been described by Cope and Marsh (*Stegosaurus*). These are the characteristics that give a realistic dimension to Doyle's novel compared with its predecessors and that determined its subsequent great influence.

It seems likely that the discovery of dinosaur footprints in Sussex, a place close to his home, suggested the idea to the British writer. Doyle placed some of these ichnofossils (casts of dinosaur tracks) in his billiard room, and he was probably fascinated by the idea of the survival of the dinosaurs, as much from the hunting perspective as from the zoologic point of view. (In fact, the main characters in his novel are two scientists, a hunter-adventurer, and a journalist.) In May 1909, Doyle wrote to Arthur Smith Woodward (1864–1944), a paleontologist at the British Museum. In his letter, which included a drawing of one of the footprints, he made a request for a specialist to visit the ichnologic sites in Sussex. This initiative clearly confirms Doyle's interest in paleontology and in dinosaurs in particular. There can be no doubt that the *Iguanodon* tracks from Sussex were the inspiration for the British author's description of the expedition members' discovery of the fresh footprints of an enormous dinosaur. Otherwise, there is evidence from letters that it was Edwin Ray Lankaster's (1847–1929) book *Extinct Animals* that directly inspired Doyle and provided the factual foundation of his novel. The choice of location for the lost world owes its inspiration to other sources.

In a 1901 paper on the osteology of *Diplodocus*, J. B. Hatcher (1861–1904), then director of the recently created Carnegie Museum (Pittsburgh), described the environment in which the giant sauropods must have lived:

> It is not improbable that during the period when these huge dinosaurs lived and flourished over what is now New Mexico, Colorado, Wyoming, Montana, and the Dakotas there prevailed throughout this region physical conditions somewhat similar to those which exist to-day in tropical America and more especially over the coastal plain of the lower Amazon with its Brazilian provinces of Amazonas and Matto Grosso with their numerous lakes and large rivers surrounded by a dense tropical vegetation with broad, flat valleys subject to periodic inundations.

It is probable that such opinions, or others similar to them, were an influence on Doyle during the creation of his novel. This is the scene described by the British novelist, but evidently, the creator of Sherlock Holmes needed some explanation for the physical isolation that allowed his lost world to exist. To this end, Doyle conceived of his fragment of

Mesozoic space on a plateau of high and inaccessible crags. It is likely that this idea was suggested to the English novelist by the explorer P. H. Fawcett, who while in London showed him photographs of such mesetas in the Brazilian area of the hills of Ricardo Franco.

The original story of *The Lost World* first appeared as a series in *The Strand Magazine*. It seems that the source of the narrative was the direct result of a bet. In a discussion, a friend of Doyle's argued about the impossibility of creating new types of adventure novels (e.g., pirates, quests for treasure). Doyle maintained that new possibilities existed that must combine imagination and realism, and he demonstrated his contention by writing a novel. Thus was born the story of the lost world. It is worth noting that this type of novel (which, as we have seen, had its precedents), does not involve a general structure different from those about treasure hunts. The only difference is that instead of gold and diamonds, the adventurers seek a possession whose value had been increasing since the beginning of the century: scientific knowledge.

The profound influence of *The Lost World* is obvious, not only on all manner of fictional stories, but also in the dreams of many paleontologists. In terms of generating the myth, it signifies, in its broadest sense, the idea that dinosaurs may have survived to the present day in some unknown part of the planet. The mythologic structure of the lost world will be analyzed later on, when I consider the procedures that have served to bring humans and dinosaurs together.

The other phenomenon of the first decades of the 20th century that I must consider is what may be called "prehistoric cinema." From 1912 to 1920, at least 13 films of this genre were made. On various occasions, this prehistoric cinema was a naturalistic attempt to show the life of our "cave-dwelling ancestors," translating current ethical conflicts to the era of the dawn of humankind. This is the case with two of the best-known titles, *Man's Genesis* (1912) and *The Primitive Man* (1913), both directed by one of the founders of cinema, D. W. Griffith. Other films in this genre were comedies that translate something other than moral conflicts to our prehistory—Charlie Chaplin in bowler hat and animal skins in *His Prehistoric Past* (1914) or as the postman who delivers stone postcards in *10,000 Years B.C.* (1917). Other films in this genre are particularly attractive in that the real protagonists of the stories are not the primitive human beings but rather the dinosaurs and other creatures of the past. In this context, the work of Willis O'Brien (1886–1962), the special effects technician, deserves special mention. He recreated *real* dinosaurs for the first time in films such as *The Dinosaur and the Missing Link* (1917) and *The Ghost of Slumber Mountain* (1919).

One of the story boards from the never-filmed movie *Creation,* a tale by Willis O'Brien and Harry O. Hoyt. Photograph courtesy of D. F. Glut.

The wide availability of these prehistoric cinema films in the United States over a relatively short period of time indicates an unquestionable interest on the part of consumers. Many of the films were based on the evolutionary theories of Charles Darwin, and it seems likely that the scandals provoked in U.S. society by the introduction of this line of scientific thought ensured their success. In any case, this genre was probably responsible for some of the most widespread prehistoric topics in the cinema and in comics—for example, the way that cavemen courted their women, whom they conquered by thumping them and then dragging them off by the hair. As far as dinosaurs themselves go, the most common theme was their simultaneous existence with primitive human beings. The origin of this supposition was probably due to *The Lost World.* The shortest time period between the appearance of humans and the disappearance of most of the dinosaurs is estimated as being more than 60 million years. Dinosaur remains had already been fossilized for millions of years by the time our oldest ancestors appeared. Without a doubt, this matter did not go unnoticed by the U.S. filmmakers of the period, who probably sacrificed scientific rigor to spectacle in order to present these enormous beasts interacting with human beings on the big screen. The contemporaneity of humans and dinosaurs has played a decisive role in the development of these myths, especially in films and in comics. There are various significant films, such as _One Million B.C._ (1940), its famous Hammer Films remake _One Million Years B.C._ (1966), and _When Dinosaurs Ruled the Earth_ (1969). More than mere spectacle, this synchronous existence of human and dinosaur introduced new elements into the dinosaur mythology, such as the humans' dependence on (and subordination to) the great beasts, and even variations on the myth of beauty and the beast, as I will examine later.

One of the most thoroughly developed phenomena in the sociocultural impact of dinosaurs in recent years—their treatment as pets—has its origin at the beginning of the 20th century, at the same time as _The Lost World_ and silent prehistoric cinema. This phenomenon is currently attaining special relevance in animated films and toys. The American artist Winsor McCay (1889–1934), author of _Little Nemo_, was one of the most important strip cartoonists of his age and was largely responsible for introducing a meticulous realism within a surreal world

U.S. poster for the first version of _The Lost World_ (First National, 1925), directed by Harry O. Hoyt. Willis O'Brien already had sequences ready for the film featuring dinosaurs in 1924, which were praised by Arthur Conan Doyle himself.

into the seventh art. In 1914, McCay created *Gertie the Dinosaur*, one of the first animated films in the history of cinema. The film begins with McCay and a group of friends stopping their car in front of the door of a museum. As they stand next to the skeleton of a brontosaur, the artist suggests to his companions that he make a film about what the enormous animal would have been like in real life. Six months later, during dinner, McCay skillfully draws the dinosaur and invites her to introduce herself. Gertie shyly pokes her head out of a cave, then emerges completely, and while coming into the midshot that she occupies for most of the film, she swallows a rock and a tree. A sea snake is terrified by the dinosaur, who, strangely, adopts the voice of her creator, successively raising her front left and right legs and crying when prompted by McCay. Gertie throws a small mammoth into a lake and then drinks the lake dry. Finally, McCay returns to the screen and is gently picked up by the brontosaur, and they both disappear.

It is obviously significant that a relationship is portrayed between a modern-day man and a dinosaur as early as 1914. McCoy gave Gertie a sweet and innocent nature, which resembled that of a child, turning a dinosaur into a pet for the first time. The Gertie phenomenon is confirmation of the special importance of dinosaurs in popular U.S. culture at the beginning of the century. The appearance of a similar phenomenon in Europe is unthinkable, even though Émile Cohl (1857–1938) had made the first animated cartoons there some years before. The understanding between humans and dinosaurs has given rise to one of the most original areas of myth, a tradition that has led to present-day animated cartoon television series such as *Dinoriders* (1988) and *Dinosaucers* (1987).

The Second Dinosaur Rush

The tremendous personal rivalry between Cope and Marsh, which on occasion ended in open disputes, was without doubt one of the reasons they were so active. The conflict was so great that it has been termed the *Great Fossil War*. In 1890, their last clashes ended as a public scandal, which made its way into the national press. This scandal annoyed members of the scientific community and the public at large and seriously limited the careers of both paleontologists until their deaths. Another probable consequence of the scandal was a cooling of interest in dinosaurs and paleontology in general.

Nevertheless, the beginning of the 20th century saw the second dinosaur rush. Various institutions pooled their efforts in the search for and exploitation of new dinosaur sites. Two of the museums that participated most actively were the Carnegie Museum (Pittsburgh) and the American Museum of Natural History (New York). This second rush generated an enormous volume of material and information about North American dinosaurs and partially contributed to the impact of the myth as sketched in the previous chapter, even taking into account the natural lag time between obtaining scientific information and its establishment in the mass media.

It is said that the U.S. patron of paleontology, Andrew Carnegie (1835–1919), commented on seeing the new great hall of the museum that bares his name: "Fill this room with something big." In attempting to fulfill this aim, Earl Douglas and William J. Holland found some new sites in the famous Morrison Formation (Upper Jurassic, 150 million years ago) in 1909. Between them, they discovered what is now one of the best-known places in the world, today known as the Dinosaur National Monument, in Utah. The Carnegie Museum's campaigns in Wyoming and Utah yielded 8 of the 13 examples of sauropod dinosaurs (excluding replicas) that may be found on display in U.S. institutions. Between 1900 and 1901, the paleontologist E. S. Riggs, from what is now the Field Museum of Natural History (Chicago), discovered various sites in the area of Grand Junction (Colorado). One of Riggs's most spectacular discoveries was the remains of an enormous sauropod dinosaur, which was given the name *Brachiosaurus*. The work carried out by paleontologists such as Douglas and Riggs represents the origin of a relatively common phenomenon in certain areas of the western United States—public awareness that urban centers are situated on the ancient *land of the dinosaurs*. This awareness results in a high degree of fusion between cultural identity and the dinosaurs. Of course, the fact that fossil sites are tourist attractions also plays its part. This is the case, for example, in the area known as the "Dinosaur Triangle," whose apexes are located in Grand Junction (Colorado), Price, and Vernal (Utah).

The paleontologic activity that contributed most significantly to the construction of the myth during the second dinosaur rush was undertaken by the American Museum of Natural History, directed by one of the best-known U.S. vertebrate paleontologists, Henry Fairfield Osborn (1857–1935). This New York institute's first studies of dinosaurs centered on Wyoming, with the excavation of now-famous places such as Bone Cabin Quarry and Como Bluff. The person mainly responsible for the American Museum of Natural History's extensive collections of dinosaur bones, which amount to many tons, was Barnum Brown (1873–1963). He was probably the dinosaurologist with the greatest fieldwork experience of all time. Brown conducted his campaigns at a multitude of sites in the western United States, Canada, South America, Ethiopia, and even India. In 1933, the preliminary fieldwork was carried out at the famous site of Howe Ranch (at the foot of the Bighorn Mountains, Wyoming). Brown obtained the patronage of the Sinclair Oil Company, which used a *Diplodocus* as its logo. The Howe Ranch sites (popularly known as "the great dinosaur graveyard") yielded more than 4,000 bones, which were sent to New York. The relationship between the Sinclair Oil Company and Mr. Bones, as Brown was known in the U.S. press, was mutually beneficial. During the 1930s and 1940s, the oil company gave away pictures and leaflets about dinosaurs at gas stations, overseen by Brown himself.

The enormous cultural impact of Barnum Brown may be inferred from several diverse circumstances. His second wife, Lilian, turned his paleontologic adventures into novels: *I Married a Dinosaur* (1950) and

Dinosaurs are a constant presence in the streets of Vernal, Utah.

Ceratosaurus Circle in the town of Dinosaur, Colorado. This town is adjacent to Dinosaur National Monument, and all of the streets are named after dinosaurs.

Bring 'Em Back Petrified (1956). Additionally, at one point in the comedy film *Bringing Up Baby* (1938), the dinosaurologist, Dr. Huxley (a character played by Cary Grant), takes on the nickname Dr. Bone.

The Sinclair Oil Company funded a project to create the life-size dinosaurs that were exhibited in the Century of Progress Exhibition at the 1933–1934 Chicago World Fair. In the exhibition, after passing through an artificial cave, the spectators arrived at a reconstruction of the Mesozoic era in which dinosaurs such as *Brontosaurus*, *Stegosaurus*, and *Protoceratops* astonished the visitors. The exhibition included a struggle between *Tyrannosaurus* and *Triceratops* in a scene made all the more dramatic by the models' ability to move their heads in a threatening manner. The fruitful relationship between Brown and the Sinclair Oil Company was renewed at the beginning of the 1960s. Once again, the oil company sponsored the dinosaur exhibit at the New York World's Fair (1964–1965). The models were created by Paul Jonas, who attempted to achieve the closest possible agreement with the known paleontologic facts of the times. The scientific consultancy was provided by Brown and John H. Ostrom of the Peabody Museum (Yale Univer-

Dinosaur memorabilia, primarily from the Sinclair Oil Company. This company financed several expeditions of Barnum Brown, the greatest field paleontologist of the 20th century. From an exhibit at the Academy of Natural Sciences in the 1980s.

sity). The models were made of fiberglass, and after the exhibition, they were donated to various parks in the United States.

One of the findings that would decisively shape dinosaur mythology was made in central Asia by expeditionary teams of the American Museum of Natural History, a long way from Barnum Brown's habitual hunting grounds. The initial reasons for mounting such a costly campaign in the search for fossils seem fairly clear, but they have little to do with the final results. Around the beginning of the 20th century, Osborn had suggested that the elevated regions of the large continent of Asia must have been the place where humans and various groups of mammals originated. The proposal was based on the belief that the most advanced forms of vertebrates, with humans at the summit, must have originated in "healthy" regions, with fresh air, in contrast to putrid, marshy areas, where only "creepy crawlies" lived. Furthermore, Osborn had the reputation within evolutionary paleontology of supporting alternative hypotheses to those of Darwinism. According to Osborn, evolution was directed by a set of genes (aristogenes) that prospectively controlled the direction of organic change; humanity had not had simian ancestors but instead was an aristocrat of the animal kingdom. Humankind's origin must be related to the superior "Aryan" race. Black ethnic groups were regressive, races that had degenerated from the original human beings. For this reason, there was no need to search for the fossil remains of the first humans in Africa.

There was almost no real fossil evidence from central Asia. The only known item was a rhinoceros molar found between Kalgan and Ulan Bator by the Russian geologist V. A. Obruchev (who, incidentally, years later wrote a story, *Plutonia*, based on the lost world myth). Starting out with such a weak scientific hypothesis, albeit one that was undoubtedly popular at the time, and above all, with so little physical evidence to back it up, it is surprising that the American Museum of Natural History threw itself into undertaking the Asian expedition so wholeheartedly.

I cannot avoid the temptation of speculating about the other decisive factors that do not figure as such in the orthodox history of the campaigns. By the second decade of the 20th century, the Wild West had been practically colonized, the Native Americans were confined to their reservations, the buffalo had been decimated, and the cowboys and legendary gunfighters had been dead for some time. The opportunities for reviving the romantic adventure of the search for fossils in American soil were scarce. Given this background, it is possible that consciously or unconsciously, the authorities of the American Museum of Natural History tried to revive the old paleontologic epic in what was in those days a large area of *terra incognita*: central Asia.

Nevertheless, the campaigns in the Gobi Desert were mounted between 1922 and 1930 and provided no small number of surprises. Osborn entrusted the leadership of the expeditions to Roy Chapman Andrews (it has been suggested that Andrews may have been the real person on whom the fictional archeologist Indiana Jones was based). Andrews and his team had to face all manner of risks, from substituting the motor oil of their vehicles with lamb fat to shooting at bandits while on horseback. In the end, the findings of the expedition had nothing to do with hominid fossils or Cenozoic mammals, but rather with dinosaur fauna and Upper Cretaceous mammals. The dinosaurs found were different from the equivalent fauna known from the United States. Among them, one of the most important types was *Protoceratops*, a small ceratopsian, dozens of whose complete skeletons were found, from newborn individuals to adults (probably of both sexes). This circumstance meant that it was possible to understand the pattern of growth of this small ceratopsian dinosaur. On the other hand, entire clutches of eggs attributed to *Protoceratops* were found that permitted the nesting behavior of this dinosaur to be inferred. Finally, remains were found on top of one of the nests of a theropod dinosaur that Osborn significantly named *Oviraptor*. The skull of *Oviraptor* was found a mere 10 cm away from an egg, which "immediately put the animal under suspicion of having been overtaken by a sandstorm in the very act of robbing the dinosaur egg nest" (Osborn, 1924).

In this way, for the first time, precise biologic information was available about how a dinosaur grew, its egg-laying behavior, what its eggs were like, and even what its possible predators were. Much of this information was published as an exclusive in several U.S. newspapers and gave rise to a notable state of expectation among the public. From the point of view of the development of the dinosaur myth, this phenomenon seems to have a clear significance: the installation in the collective consciousness of the idea that dinosaurs were definitely living beings, with their own particular habits, just like those of the multitude of *real* (present-day) animals.

It should be said that we currently know that the nests and eggs originally attributed to *Protoceratops* are really those of oviraptosaurs. *Oviraptor*, which has been confused with an egg robber, is in reality responsible for the famous eggs laid in Mongolia during the Upper Cretaceous. Their relation with the nests needs reinterpretation. Probably they would have defended their eggs, shaded them, or even controlled their temperature. In any case, this new evidence does not alter the great relevance that these discoveries from the 1920s had on the social and cultural impact of dinosaur biology.

The spread of information about the study of dinosaurs among the public through attractive lifelike reconstructions had been a particular concern of Osborn's since the first years of the 20th century. For this reason, the collaboration between Osborn, paleontologist William Diller Matthew (1871–1930), and artist Charles R. Knight (1874–1953) was of vital importance for the sociocultural advance of the dinosaurs. The work of this group not only produced a multitude of reconstructions of dinosaurs and other animals of the past, but also greatly influenced the techniques for exhibiting fossil skeletons. This was a new conception of museology in which skeletons were presented mounted in lifelike postures that were judged to be typical of the animal in question. Adding these conceptions to the fact that an enormous number of complete dinosaur specimens had been discovered, we can understand the reason why this period of history was characterized by the public's massive attendance at museums. The enormous specimens of sauropods, stegosaurs, and ceratopsians became the main exhibits, and there is no doubt that they attracted more visitors.

Charles R. Knight is one of the most respected and imitated animal reconstructive artists. His work, professional career, and manner of working are the paradigm of this particular aspect of the popularization of natural history. He applied his profound knowledge of present-day animals (especially their behavior, osteology, myology, and external morphology) to the job of visualizing animals, such as the dinosaurs, that nobody had ever seen in real life. During the final decade of the 19th century, Knight began to collaborate actively in the field of animal illustration for children's books and magazines. To increase his knowledge, he would go to the zoo and the department of taxidermy of the American Museum of Natural History, where he began to become well known. In 1894, he dazzled the paleontologists with his lifelike reconstruction of the mammal *Elotherium* from its skeleton, which attracted Osborn's attention. His close collaboration with the dinosaurologists was not limited to Osborn. Some of his most spectacular works were the fruit of several months' visit to Philadelphia in the company of E. D. Cope. Knight's work can be admired in many books, such as *Before the Dawn of History* (1935) and *Life through the Ages* (1946). His museum work can still be seen in the Natural History Museum of Los Angeles County, Chicago's Field Museum, and the American Museum of Natural History in New York. Knight's work was decisive in the generation of the myth. His dinosaur reconstructions have been repeatedly copied (in a decidedly open way on many occasions) in a multitude of pictures and comics, and they have had a tremendously broad influence in the world of cinema.

At the beginning of the 20th century, in this atmosphere in which dinosaurs were considered as living things, rather than the static symbol represented by bones within rock sediments, the filmmaker Willis O'Brien undertook his work. O'Brien is normally credited with being the creator of the techniques of stop-motion animation. By using this procedure (i.e., models animated frame by frame), O'Brien made the first more or less realistic forays into putting dinosaurs on the big screen. Since he had been a boy, O'Brien had been fascinated by prehistoric life. At the age of 16, in 1902, he was taken on as a guide with a University of Southern California fossil expedition to the Crater Lake area. In 1917, the Edison film company, for whom O'Brien worked, tackled a series of weekly educational films that included short films about prehistoric life. Until that moment, O'Brien had created his creatures by

The *"Brontosaurus"* from *The Lost World* (First National, 1925).

An *Allosaurus* (which lived in the Jurassic Period) attacks a *"Trachodon"* (a duckbill dinosaur, now known as *Edmontosaurus,* from the Cretaceous Period) in the film *The Lost World* (First National, 1925).

basing them on popular reconstructions, but now he decided that he needed professional advice. To this end, he decided to consult Barnum Brown. Thus armed with the necessary scientific information and an increasingly polished animation technique, O'Brien became one of the most widely recognized special effects technicians. In 1919, he was hired by W. R. Rothacker, who had the idea of transferring a version of Doyle's novel, *The Lost World*, to the cinema. By 1923, they had completed all the preproduction work, including the building of 50 models of dinosaurs on various scales. This task was undertaken by the Mexican artist Marcel Delgado, who was inspired by the murals of Charles R. Knight. Delgado's work was so meticulous that he included a rubber bladder inside the models to simulate the animals' breathing.

The Lost World opened in 1925 and rapidly became well known and highly popular. (The film has been restored and rereleased on DVD.) The structure of the story is similar to that of the novel, but the treatment of the dinosaurs is different. They appear only sporadically in the book, but they are the real stars of the film. Many more forms from the U.S. dinosaur fossil record made their appearance: *Apatosaurus,* *"Anatosaurus," Allosaurus, Triceratops,* and *Styracosaurus.* The various behaviors of these dinosaurs, which in some cases were generated from the scientific interpretations of the times, are of particular note. A good example of this is the case of the care of the offspring by a *Triceratops,* probably a female, who actively defends its young against the attack of a carnivore and even licks them tenderly. Other behaviors seen in *The Lost World* seem to have different explanations, like the threat display of an *Apatosaurus,* which shows its powerful teeth, even though Professor Challenger, the protagonist, considered it to be a peaceful herbivore. It is likely that this phenomenon was a response to the need to give the dinosaurs some kind of personality by using the resource of a threat display used by many carnivorous mammals that could be quickly identified by the audience. In fact, O'Brien resorted to this again in the case of the amphibious *Apatosaurus* that actively pursues a man in *King Kong* (1933). Finally, the appearance of a herd of ceratopsians (dino-

The Beast of Hollow Mountain (United Artists, 1956), based on a story by Willis O'Brien.

saurs with horns) in the first filmed version of *The Lost World* should be mentioned. The gregarious character of these dinosaurs was probably inferred from the discoveries of *Protoceratops* in Mongolia. Neverthe- less, the appearance of monogeneric bone beds (of *Styracosaurus* and *Centrosaurus*, etc.) in the Upper Cretaceous of Alberta (Canada) has only recently been proposed on the basis of a seemingly more conclu- sive type of evidence. We will find out more about these gregarious behaviors in a wide range of dinosaurs later.

5

Dinosaur Cartoons: *Fantasia*

Shortly after W. McCay, the New York animator John R. Bray made a new version of *Gertie the Dinosaur*, and in 1915 another animated film about dinosaurs for Pathé, *A Stone Age Adventure*. In 1925, the famous character Felix the Cat journeyed into the past in *Felix the Cat Trifles with Time*, which featured for the first time some of the gags now common in dinosaur cartoons, such as Felix confusing the back of a brontosaur for stones on which he tries to cross a lake. The stories of many of these dinosaur cartoons are similar to those of the prehistoric cinema I have already mentioned. The main character is a primitive human or a humanized animal; the anthropomorphization of dinosaurs in comics and animated films comes from this period, as does the comic use of particular dinosaur structures, such as spines as stairs or the teeth of a saw.

Between 1938 and 1941, the Fleischer brothers made a series of dinosaur cartoons for Paramount. In *Granite Hotel* (1940), a sauropod turns into a prehistoric firefighters' vehicle. In *The Arctic Giant* (1942), Superman comes face to face in Metropolis with a tyrannosaur preserved in the ice of Siberia. The struggle causes chaos and the destruction of most of the city. *The Arctic Giant* is an early instance of the "dinosaurs versus civilization" theme, which I will deal with later. Nevertheless, we should remember that the final sequences of *The Lost World* (1925) show an enormous sauropod in the streets of London destroying buildings and terrorizing the population.

Fantasia, a risky bet by the Walt Disney studios, premiered in 1940. Various symphonic works served as a basis for the cartoons. The score of Igor Stravinsky's *The Rite of Spring* was translated into images that recount the history of the planet from before the appearance of life and the origin of the first unicellular organisms up to the disappearance of the dinosaurs. The story has a distinctly evolutionist feel, which caused the film some problems with U.S. ecclesiastical authorities. Beyond even that, Disney's original idea was to continue to tell the story of life up to the origin of humanity, but this was made an impossible task as a result of the threats from some religious fundamentalists.

One of Disney's first concerns was to make the story of the history of living beings a serious one, the fruits of a meticulous faithfulness to the scientific knowledge of the time. In fact, the creator of Donald Duck asked his cartoonists to animate their dinosaurs with real movements, not like those of Pluto and Snow White's dwarfs. Among other things, this meant the practical elimination of one of the trademarks of the Disney factory: the anthropomorphizing of animals. In addition, the studios brought in various paleontologists as advisors: Roy Chapman Andrews, Barnum Brown, and Julian S. Huxley (1887–1975). The reconstruction of the dinosaurs in _Fantasia_ even interested the New York Academy of Science, who asked for a private screening of the film. This scientific institution thought that Disney's dinosaurs might serve the popularization of science better than the entire mass of fossil and taxidermal material in the museums.

Despite these assertions, there are certainly some difficulties with the interpretation of the dinosaurs in _Fantasia_. The way of life of the

The three-fingered _Tyrannosaurus rex_ in _Fantasia_ (Walt Disney Productions, 1940).

duck-billed dinosaurs and sauropods (amphibian forms that consume aquatic plants) is consistent with the biologic model of these dinosaurs that was held to be true at that time. On the other hand, the concurrence of primitive Permian synapsids and Jurassic and Cretaceous dinosaurs undermines the initial attempt at realism. The climax of the story is a dramatic fight between a stegosaur (Upper Jurassic) and a tyrannosaur (Upper Cretaceous) with bloodred eyes. The two animals are separated in time by about 80 million years. This visually strong sequence, illustrated the Darwinian concept of the "survival of the fittest," was described as follows by J. Culhane: "*Stegosaurus* was more than twice as big as an elephant and wore a heavy coat of scales, almost like a suit of armor. There were four huge spikes at the tip of his powerful tail. He is attacked by *Tyrannosaurus Rex*—King of the Tyrant Lizards—a dinosaur standing eighteen feet tall and measuring about forty-seven feet overall. With his enormous jaws, lined with saber-sharp teeth six inches long, he brings *Stegosaurus* down and breaks his neck" (*Walt Disney's "Fantasia,"* 1978). These sentences accurately describe the type of approach that the Disney studios wanted to adopt for their dinosaurs. In truth, the entire episode is filled with constant dramatic reminders of the relationship between predator and prey. A *Pteranodon* catches a fish, and both are trapped in flight in the enormous jaws of a mosasaur. A small carnivore tries to trap an *Archaeopteryx*; a group of ceratosaurs harass the stegosaurs; and so on.

The Disney conception of dinosaurs is a misinterpretation of Darwinist hypotheses and is much more closely aligned with social Darwinism than with the proposals of the great British naturalist. The possible relationship between dinosaurs and social Darwinism does not seem obvious, but it jibes with ideas that were popular among captains of industry during the first third of the 20th century. One of the industrial champions of social Darwinism was Andrew Carnegie, who considered that the great dinosaurs (like his own species of *Diplodocus, D. carnegii*) were obvious proof of the great capacity of the evolutionary process. Carnegie was largely responsible for the popularity of *Diplodocus* because he gave plaster copies of this North American Jurassic sauropod to various institutions around the world, including London, Paris, Berlin, Madrid, Vienna, México, and La Plata.

Fantasia (1940) described the extinction of the dinosaurs in a particular way: the dinosaurs died of thirst in a world of rising temperatures. This interpretation comes from one of Barnum Brown's suggestions concerning the famous Howe Ranch dinosaur cemetery: the great lakes dried up, the marshes disappeared, and the dinosaurs had to crowd

Replica of the skeleton of *Diplodocus carnegii* that the multimillionaire "King of Steel," Andrew Carnegie, donated to Madrid's Museo Nacional de Ciencias Naturales. For many years, this enormous sauropod has been the only assembled dinosaur skeleton in Spain. Courtesy of the Museo Nacional de Ciencias Naturales, Madrid.

together in the small bodies of water. Although Barnum Brown's hypothesis referred to a local phenomenon, the Disney studios gave it a planetwide dimension, contributing to the popular conception that dinosaurs disappeared rapidly from a dry and dusty Earth.

All things considered, *Fantasia* is one of the cinematic works that has most contributed to the mythologizing of the dinosaurs. The film was largely responsible for establishing in the popular imagination the singular conception of the dinosaur as both pathetic, tragic victims and ruthless killer predators. This characteristic, combined with the idea of the extinction of dinosaurs in a hot environment, can even be traced to recent films, such as *The Land before Time* (1988).

6

Were Dinosaurs Stupid?

The dinosaurs in *Fantasia* (1940) were interpreted according to the particular paleobiologic model prevailing at that time. This model, which considered dinosaurs to be slow, heavy, and stupid animals, large lumps of meat endowed with a vegetative life, was basically supported by three lines of argument, as follows.

1. The brain capacity of dinosaurs was considered to be ridiculously small relative to their enormous body size. To correct this supposed deficiency, some dinosaurs, such as the stegosaur, would have developed another brain at the rear in the sacropelvic region. This idea was born out of an inadequate interpretation of the broadening of the medullar cavity in the sacral region. It is currently believed that this cavity does not represent a thickening of the nervous structures, but rather was a center for the storage of glycogen, which is of great importance in sugar metabolism. The idea of dinosaurs with two brains was popular nevertheless, especially in the 1940s and 1950s. In the Japanese film *Gojira no Gyakushyu* (1955) (entitled *Gigantis, the Fire Monster* in the United States), two paleontologists reported the characteristics of the monsters Godzilla and Anzyllas to the army (see chapter 23). The latter, bearing a distant resemblance to an ankylosaur, is said to have three brains: the normal one, another below its chest, and another in its abdomen.

2. A second supposed line of evidence for considering dinosaurs as clumsy and stupid beasts was the interpretation that their metabolism was equivalent to that of present-day reptiles (despite the obvious differences, such as the unequal amount of biomass that they must give rise to).

3. The idea of gigantism itself has traditionally been associated with a congenital lack of intelligence. It is a fact that there are innumerable stories from all periods that tell of giants, ogres, and cyclops who are fooled by the astuteness of human beings.

There is no doubt that dinosaurs were considered to be huge terrestrial animals with slow and heavy movements and mechanical reactions. Later, the reaction against these interpretations formed the basis for the new conceptions of the currently prevailing model, which arose at the end of the 1960s.

In addition, many dinosaurs had developed strange structures—the dermatoskeletal plates of the stegosaurs or the cranial projections of the duck-billed dinosaurs—that were difficult for the paleontologists of that time to explain functionally (i.e., adaptively). Everything seemed to indicate that the dinosaurs were failed experiments of a creative Nature and that they had no option but to die out.

This interpretation of dinosaurs in the general context of the story of life definitely comes from earlier times but was clearly expressed during one of the moments of greatest development of the evolutionary hypotheses in what was known as the theory of orthogenesis. This theory held that most linear evolutionary change came about independently of environmental conditions. In 1924, Richard Swan Lull interpreted the extinction of the dinosaurs in the light of the orthogenetic suppositions developed by another U.S. paleontologist, Alfeus Hyatt, according to whom any organic group passes through stages similar to those of an individual: youth, maturity, senility, and death (extinction). Lull suggested that this "racial senility" was responsible for the disappearance of the dinosaurs. At any rate, Lull did not think of the dinosaurs as "useless experiments" in the history of life. In 1933, he argued that the dinosaurs did not represent a futile attempt by Nature to populate the world with creatures of momentary insignificance but rather were comparable in their majestic expansion, slow culmination, and dramatic fall to the great nations of antiquity.

Independent of the unique orthogenetic flavor of these words, there arises another idea that constituted a clear conflict between various dinosaurologists of those decades. In the face of the dinosaurs' inexorable fate, the consequence of their ultimate incapacity to adapt, there is an almost reverent recognition of their extraordinary longevity of more than 100 million years. The interpretative model of the inexorable destiny of the dinosaurs on their way to extinction reappears in the scientific and popular literature even at the end of the 1960s. In 1970, W. E. Swinton made reference to the "racial old age (phylogeronty), lack of plasticity, both physical and mental, of the dinosaurs" to explain their extinction, at least partially.

The model of dinosaurs as stupid and heavy had wide repercussions in the mass media. In 1956, the important U.S. science fiction

The Giant Behemoth (Allied Artists, 1959). This was the first dinosaur to menace London after the sauropod in *The Lost World* (First National, 1925).

writer and essayist L. Sprague de Camp (1907–2000) published his story "A Gun for Dinosaur." The enormous costs of the upkeep of a time machine means that their services have to be lent to a syndicate of hunting guides who organize trips into the remote past. The story tells of the tragic end of a safari at the beginning of the Upper Cretaceous. After firing at an *Ornithomimus* and killing a pachycephalosaur for meat, the safari comes across a large theropod, which eats one of the hunters. Two assessments of de Camp's story are typical of the stupid dinosaur model. Big-game hunters are used to killing mammals and tend to shoot them in the head, with the aim of injuring their brain. Given their size, this approach is almost useless in the case of dinosaurs. For de Camp, the majority of dinosaurs, in spite of being stupid, have a good sense of smell, keen vision, and fairly sensitive hearing. They lack intelligence, and their primary weakness consists of having no memory. For this reason, when they stop seeing an object, they forget about it, and so it is easy to escape their pursuit by hiding. A similar idea is put forward in the book *The Dechronization of Sam Magruder*, written by the famous U.S. paleobiologist George Gaylord Simpson between 1970 and 1984 (the year of his death). The manuscript was discovered by his daughter Joan Simpson Burns and was published in 1996. The story tells of the fight for survival of a chronologist (Sam Magruder) from the 22nd cen-

Illustration by the Spanish artist Mauricio Antón for Poul Anderson's story "Wildcat" (1964). An enormous theropod dinosaur remains active even after it has been eviscerated by the bullets of a heavy machine gun.

tury, who has been accidentally transported to the Upper Cretaceous while investigating the properties of time. Magruder ratifies, through direct field observation, the stupidity of dinosaurs, in this case the great carnivores *Tyrannosaurus* and *Gorgosaurus*. If the prey hides from the sight of the dinosaur, the predator quickly forgets about it. Dinosaurs do not have memories.

Poul Anderson's story "Wildcat" appeared in 1958. A U.S. oil company maintains a base for oil exploitation in the Upper Jurassic. The crude oil is sent to the present time. The humans of the base have to remain on permanent alert, and the clashes with dinosaurs are constant:

> It was really a tyrannosaur. . . . It staggered up to them with the excessive weight and clumsiness the paleontologists had spoken of. Some regarded it as having been a gigantic species of hyena, a scavenger. They had forgotten that, like the snake or crocodile of the Cenozoic, it was not intelligent enough to recognize that dead meat was a possible source of food and that the brontosaurs on which they fed were even more stupid.

Anderson's daring ideas are based on the strict inclusion of the dinosaurs within the level of organization of present-day reptiles, with a generous supplementary dose of misinformation. Anderson supposed that dinosaurs would have a tremendous vitality against wounds and mutilations ("a reptile dies with greater difficulty, since it is less alive"), describing in a terrifying fashion the encounter between several off-road vehicles and a giant theropod.

The stupid dinosaur model, of course, does not allow for any type of behavior that cannot be seen in present-day reptile forms. The orthodoxy made a strong attack on *Triceratops*' tender parental care in *The Lost World* (1925) and other similar manifestations. In 1964, the U.S. dinosaurologist G. L. Jepsen affirmed that the pastoral scene in paintings and films that depict individuals of a *Triceratops* clan, peacefully distributed in small family groups of a father, mother, and offspring, was thus a picturesque superstition.

This state of affairs was set to change at the end of the 1960s and beginning of the 1970s with the appearance of a new interpretative model of dinosaurs, known as the Dinosaur Renaissance.

7

Dinosaur Renaissance

Around the mid-1960s, various paleontologists, working in different fields, began to come up with a new model for the paleobiologic interpretation of dinosaurs that substantially modified the old approaches and launched the Dinosaur Renaissance. One of these specialists was Robert T. Bakker, who had championed the most vigorous heterodoxy in his interpretation of dinosaurs. One of the keys to these new attitudes was the heat-regulation regime of the dinosaurs. Bakker tried to demonstrate their endothermic (hot-blooded) nature by a series of observations about the proportions of predators and prey in continental fossil vertebrate communities. A smaller percentage of endothermic predators, which have a large energy requirement, are tolerated than of exothermic (cold-blooded) predators. According to Bakker, the percentage of predators in the dinosaur communities was similar to that of mammals, thus implying that they were endothermic. In 1969, the U.S. dinosaurologist John H. Ostrom described the new carnivorous dinosaur *Deinonychus*, a medium-sized form (some 3 m in length) that seemed to show evidence of a great capacity for coordination and sense of balance that were hard to associate with a typical present-day reptile. From the beginning of the 1970s, Ostrom himself relaunched the idea that birds arose from dinosaurs, these flying vertebrates being definitely endothermic. (It was Ostrom who resurrected the idea of dinosaur endothermy in professional circles, but it was Bakker's 1975 article in *Scientific American* that brought the idea to the public at large.) In 1974, Bakker and Peter M. Galton proposed the creation of a new class of vertebrate, "Dinosauria," which would include all groups of dinosaurs and birds. This proposal, which was perfectly consistent with the phylogeny, was deplorably attacked in its time, even by some specialists who defended the relatedness of dinosaurs and birds.

Dinosaur artist Gregory S. Paul drew this whimsical picture of a world with both humans and dinosaurs. © Gregory S. Paul.

At the end of the 1960s, the French paleohistologist A. de Ricqlès also argued in favor of the supposed endothermic nature of dinosaurs on the basis of his study of the internal structure of dinosaur bones. Many other lines of argument have been proposed in support of or against the endothermic dinosaur model. Some of the original ideas of Bakker, Galton, Ostrom, de Ricqlès, and other paleontologists have been refuted. These days, there is a broad consensus that asserts that the dinosaurs were not typical reptiles, either in morphology or paleobiology. Some authors continue to support the hypothesis of endothermy for the entire group. Others maintain that probably only those forms that were structurally closest to birds were endothermic. The remainder could have had a certain degree of homeothermy (constant body temperature), especially those of great size. Additionally, complex patterns of behavior, typical of mammals and birds, are currently attributed to dinosaurs: care of nests and offspring, and gregarious habits with sophisticated intragroup relationships among the individuals.

Particular functional morphology studies have even altered our conceptions about the external appearance that dinosaurs must have had, as well as their posture and the bodily disposition of their walk. In this sense, one of the suggestions that has generated the greatest controversy with respect to the old reconstructions of dinosaurs came from the U.S.-based British paleontologist Peter M. Galton in 1973. The buccal structure of mammals is configured by the appearance of a muscular wall forming the cheeks, which is absent in present-day reptiles. The ornithischian dinosaurs, the bipeds and quadrupeds, are phytophagous forms. All ornithischians, except the most primitive ones, have a lateral surface on the bones of the mouth that was probably used for the insertion of a muscular cheek, analogous to that of mammals. These cheeks would prevent the loss of food through the corners of the mouth during chewing, a structure similar to that of present-day cows and camels. From a merely iconic point of view, this interpretation of the ornithischians assumes that the mouths of these dinosaurs are no longer reconstructed with a reptilian appearance but with a mammalian one. (Recent work by Larry H. Witmer has cast doubt on the size and nature of ornithischian "cheeks.") At present, in some cases, the reconstruction and illustration of dinosaurs have stopped being exclusively an aid to the popularization of paleontology and have become a vehicle for interpretation and for furthering specific scientific proposals. This has happened above all in the case of specialists who were also draftsmen, such as Robert T. Bakker and Greg Paul. In 1987, the U.S. paleontologists Joel K. Hammond and Peter Dodson claimed that art was becoming closer to science and that the new dinosaur art no longer only illustrated, but simultaneously defined and resolved problems.

Some paleontologists have contributed in recent years to the increase in the spread of "dinosaurology" among the public, putting forward nicknames for newly discovered dinosaurs. These sobriquets are used for various purposes, among them that of rapidly identifying a new species of dinosaur until a formal name is published. This is the case of "*Ultrasaurus*," "*Supersaurus*," "Claws," and *Corretón* —the Big Runner (*Pelecanimimus*, an ornithomimosaur from the Lower Cretaceous of Cuenca, Spain). On some occasions, as is the case with *Seismosaurus*, a U.S. diplodocid, the nickname has become the formal generic name.

The press and television are essential for spreading these nicknames. On certain occasions, the mass media has run ahead of paleontologic research into dinosaurs. In 1970, Galton admitted that his ethologic interpretation of the pachycephalosaurs was inspired by reading L. Sprague de Camp's 1956 story, "A Gun for Dinosaur." The pachycephalosaurs were enigmatic bipedal ornithischians with considerable development of the bone thickness of the cranial roof, a trait that is difficult to explain from a functional point of view. De Camp supposed that this trait would be used for ritual combat by a clashing of horns, in a similar way to the behavior of present-day sheep, and this was precisely the explanation that Galton proposed. In 1985, José F. Bonaparte, from the Museo Argentino de Ciencias Naturales "Bernadino Rivadavia," described for the first time the discovery of a new form of carnivorous dinosaur from the Argentinean Cretaceous that, significantly, was called _Carnotaurus_. One of this predator's most notable traits was the presence of a pair of horns situated above the orbits, a typical characteristic of a phytophagous rather than a carnivorous animal. For this reason, its possible real existence before its discovery was unlikely. However, in 1958, Superman came across a dinosaur surprisingly similar to _Carnotaurus_, which he even fought as if it were a bull!

Research on dinosaurs has snowballed remarkably in recent decades. Between 1970 and 1987, 40% of all the forms known to date were discovered; specialist interest has specifically opted for the biology of dinosaurs, and this fascination has spread to the mass media. One of the most obvious modern expressions of this interest in the way of life of the dinosaurs is the appearance of businesses such as Dinamation and the Japanese company Kokoro. Both have sold or rented out an enormous number of life-size reconstructions of the most popular dinosaurs. The models have a special feature: they are able to make limited movements and emit supposedly natural sounds. Though first proposed by Othenio Abel in the 1920s, the idea that dinosaurs were vocal animals, able to emit sounds with the intention of communicating with other animals of the same species or even to intimidate those of other species, clearly belongs to the new conception of the dinosaur made popular by the Dinosaur Renaissance. It has been developed by several authors, especially the U.S. dinosaurologists J. A. Hopson and D. B. Weishampel, in duck-billed dinosaurs (hadrosaurs). According to both authors, the hollow cranial structures of these ornithopods would have served as resonating chambers. In the comedy _Caveman_ (1981), the same dinosaur howls at the moon at night and crows like a rooster at dawn.

Later, we will see more examples of the great impact that this currently prevailing biologic model of dinosaurs has had on the mass media. The Dinosaur Renaissance has given a more objective vision of what would have been the way of life of these animals. This model views dinosaurs not as strangers to the natural order, but as animals that are perfectly integrated within it. A carnivorous or phytophagous dinosaur would not be so distant, in ecological terms, from today's lion or elephant. This context made paved the way for the appearance of books such as Michael Crichton's _Jurassic Park_, which was published in 1990

Two *"Trachodon"* (duck-billed dinosaurs) from the classic film *The Lost World* (First National, 1925). This film has recently been restored and released on DVD (Image Entertainment, 2001).

and which hit the cinema screens in 1993. Steven Spielberg's film in particular manages to transmit in two hours the ideas that the paleontologists of the Dinosaur Renaissance had been trying to communicate to people for years.

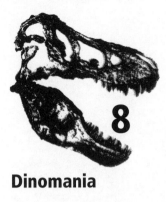

8

Dinomania

It seems beyond doubt that the phenomenon of the films *Jurassic Park* (1993) and its sequels, *The Lost World: Jurassic Park* (1997) and *Jurassic Park III* (2001) has generated a great deal of dinosaur mania, or "dinomania," in recent years. Dinomania is, however, a sociocultural phenomenon whose origins must be sought, as I have shown, in the second half of the 19th century in Victorian England, during the period of the collaboration between Richard Owen and Benjamin Waterhouse Hawkins. Since then, dinomaniac phenomena have resurged on various occasions and for the most part in the United States. In fact, in recent years, they have become a phenomenon closely associated with the development of new products in a market that is keen to acquire all manner of dinosaur-like objects, which are termed "dinomorphs." At this point, we may attempt to define dinomania as the impulse to surround oneself with all forms of dinosaur iconology and to hoard information about their appearance, size, and way of life. The phenomenon is similar to that which surrounds film, television, and sports stars.

The section of the population upon which dinomania rests is largely made up of children from 6 to 12 years of age. Children draw and pore over pictures of dinosaurs avidly while making a great show of their knowledge of the group. In 1994, a well-known contest on Spanish Television exhibited the feats of memory of a child able to retain a huge quantity of information about dinosaurs. Dinosaurs have become real childhood idols, in some cases being as highly revered as the heroes of fairy tales. In Wes Craven's film *Wes Craven's New Nightmare* (1994), a stuffed-animal *Tyrannosaurus* is the guardian in whom a child, threatened from the dream world by the infamous villain Freddy Krueger, places his trust.

There are no clear explanations for children's fascination for dinosaurs. It has been said that in the world of children, the dinosaurs play a primordial role in the development of the young. Children may learn to control their terror in the face of such frightening creatures that straddle the imaginary and real worlds because they can be certain that the dinosaurs became extinct millions of years ago. They thus act as potential exorcists of childhood fears, having an advantage over other monsters (e.g., vampires, ghosts) whose nonexistence may be questioned. However, it is clear that the current image of dinosaurs is generally devoid of this supposedly terrifying power. Notable exceptions to the general rule can be found outside the realms of the comics, cartoons, and the industrial toy trade in films such as *Jurassic Park* and *Carnosaur* (1993). It should also be noted that in the most recent films, unlike in those of years ago, only the carnivorous dinosaurs (*Tyrannosaurus rex, Dilophosaurus, Spinosaurus,* and *Velociraptor*) have assumed the role of villain.

Other reasons advanced to explain our fascination for dinosaurs include the following. (1) Human beings retain an inherited ancestral terror of our distant ancestors (primitive mammals) that lived alongside the dinosaurs. This improbable proposal was suggested in 1924 by the German paleontologist Edgar Dacque in his book *The Primitive World: Legend of Humanity.* (2) Dinosaurs make up an alternative world of terrestrial animals that stimulates people's imagination. In fact, the certainty that our present-day world, populated by lions, zebras, and elephants, was foreshadowed by a radically different and unusual community from millions of years ago could even produce a certain sense of awe, which could manifest itself by, for example, the shiver provoked by the sight of the bones of a diplodocus or stegosaur. (3) Dinosaurs have replaced the myth of dragons in modern society. This phenomenon will be examined later, along with Japanese cinema featuring giant monsters. (4) Although not all dinosaurs were large, among them we find the inhabitants of the terrestrial environment with the greatest known biomass, as in the case of many sauropods. The enormous dimensions of a brachiosaur or a tyrannosaur can cause a feeling of uneasiness in human beings, who are fascinated by their potential for power and domination.

Paleontologists are, of course, implicated in the generative structure of dinomania. The information generated by the dinosaurologists is transmitted to society, which avidly devours news of discoveries of new dinosaurs, especially if they have some type of previously unknown, spectacular characteristic. In this respect, the news of the discovery of "the biggest," "the oldest," "the biggest claws," etc., are particular causes for celebration. As a result of public demand, dinosaurologists continue their investigations, which are ever better funded in general terms. This is something that does not happen in other areas of paleontology. It is a matter of a feedback process that in recent years has generated an amazing amount of information about dinosaurs, accompanied by an authentic golden age from the sociocultural perspective.

Some paleontologists believe that the feedback may even have ramifications in the purely scientific sphere, in the sense that popular opinion may end up conditioning scientific hypotheses. Some years ago, the U.S. paleontologist Larry Martin (defender of the hypothesis that relates birds and crocodiles) suggested that the reason for the general acceptance of the dinosaur origins of birds by scientists was the result of their popularity. It is clear that the dinosaur-origin-of-birds hypothesis has been rapidly accepted, especially by dinosaurologists. During the last decade, alternative positions (such as linking dinosaurs to crocodiles) have frankly become those of a minority.

From the point of view of systematic biology, birds must be considered to be dinosaurs with feathers. I invite readers who are scandalized by this statement to compare the skeletons of a dinosaur such as _Velociraptor_ or _Dromaeosaurus_ with the oldest known bird, _Archaeopteryx_. The similarities between both skeletons are much more than mere coincidence. (John H. Ostrom's article on the subject lists 50 skeletal similarities.) They are the product of a close common ancestor. The acceptance of the proposal that birds are flying dinosaurs is increasingly widespread with the public. It means that dinosaurs did not become completely extinct; they live on as birds. For 150 million years, they represented the dominant continental vertebrate fauna. From 65 million years ago, in terrestrial environments, mammals succeeded dinosaurs, which continued to predominate in the aerial environment.

The acceptance of this hypothesis implies that dinosaurs, in their winged form, have played an important role in the history of humankind. Birds have always been considered to be symbols of peace and freedom—this being a highly relevant sociocultural role even in our age. Feathered dinosaurs even play a predominant role in the history of humankind's food. In a famous trailer, the British filmmaker Alfred Hitchcock ate a roast chicken while advertising his film _The Birds_ (1963). Now we are aware that a dinosaur was being eaten and that in Hitchcock's film, the act was avenged by other dinosaurs that we call gulls, crows, and starlings. It is also clear that most dinosaurs became extinct forever. Nevertheless, they frequently appear contemporaneously with

This card is from *Reign of Pterosaurs,* a set inspired by the many card series devoted to dinosaurs.

human beings in fantasy stories. In the following chapters, the procedures used in the fantastic discourse will be looked at with the aim of explaining the coexistence of humans and the dinosaurs that became extinct (those known as the nonavian dinosaurs).

9

The Synchrony of Humans and Dinosaurs

Any fantastic discourse in literature, cinema, or comics basically arises from the overlap of reality and unreality. As long as Count Dracula stays in his own unreal world, he will never be a part of human fantasy. This stance may be evaluated tautologically if, simplistically, unreality is defined as the negation of the realm of the real. However, in the case that concerns us—the mythology of dinosaurs—the worlds of the dinosaurs and that of human beings are both real, the only difference between them being temporal. One hundred million years ago, dinosaurs were a part of the reality of Nature. Now, only their fossil remains are real. From this point of view, the interaction between humans and dinosaurs immediately gives rise to a fantastic discourse. The dinosaurs are systematically present in the contacts between human beings and our distant past as described in fantasy tales.

The temporal overlap of humans and dinosaurs obviously has to be justified in some way. In the chapters that follow, I will examine the details of the procedures called on to do this: prehistoric cinema; the myth of the lost world; dinosaurs that are frozen or in suspended animation; time travel; dinosaurs of the future (reappearance); exodinosaurs (appearance on other planets); and the new frontiers of science (genetic engineering).

The origin of what is known as prehistoric cinema has been discussed before. In this genre, the problem of allochrony is overcome immediately: quite simply, it is denied. The contemporaneity of our distant ancestors and dinosaurs is, in fact, one of the most important aspects of films such as *One Million Years B.C.* (1966) and *When Dinosaurs Ruled the Earth* (1969). The second of the procedures mentioned above, concerning the idea of the survival of dinosaurs in some inaccessible place, is described next.

The late dinosaur paleontologist Jim Jensen of Brigham Young University "capturing" a *Camptosaurus*. Photograph courtesy of Jim Jensen, to M. Brett-Surman.

A famous dinosaur cartoon by legendary artist Charles Adams, who published many of his works in the magazine *The New Yorker.*

10

The Myth of the Lost World

The enormous importance of Arthur Conan Doyle's 1912 novel *The Lost World* and its broad repercussions in the generation of the dinosaur myth were discussed in chapter 3. The original idea has frequently been revisited, exchanging the original Brazilian plateau for hidden valleys or isolated islands. Skull Island is an example of the latter, where the giant ape King Kong (in the 1933 movie *King Kong*) lived. The famous film describes the dramatic story of an enormous simian plucked from its home and moved to New York, where it perishes (as is explained in the film)—not as a result of being brought down by the bullets of the hunter planes, but as a result of the beauty of the main character, Ann Darrow, played by the actress Fay Wray. King Kong lives in a world shared with a multitude of dinosaurs and other creatures from our remote past, including enormous carnivores such as allosaurs, stegosaurs, and sauropods, as well as a large *Pteranodon* and an unusual plesiosaur that behaves like a constrictor.

King Kong is a complex film, rich in symbols and myths (e.g., beauty and the beast, the cave, the supermale, Oedipus, the conflict between humankind and Nature, gigantism) that are situated in the subconscious. Above all, *King Kong* is a true and authentic homage to the entire surrealist movement, as André Breton argues: "I believe in the future resolution of these two apparently contradictory states, such as are dream and reality, in a type of absolute reality: surrealism. It is a question of leaving the subconscious at perfect and absolute liberty." Breton's words perfectly illustrate the dreamlike atmosphere that the film reflects on screen, in a species of grand psychoanalytical poem of the monstrous, of the abnormal, of the strange. *King Kong*'s success encouraged its makers to release a sequel, *Son of Kong*, in the same

This is the original scale model for the *Stegosaurus* in *King Kong* (RKO Pictures, 1933). It is now in a private collection. Photograph by M. Brett-Surman, courtesy of Forrey Ackerman.

year (1933). Kiko, an albino son of King Kong, battles with the dinosaurs of Skull Island in defense of the human beings.

Other scenes of insular lost worlds are presented in films such as *Unknown Island* (1948) and *Two Lost Worlds* (1950). The latter is a modest story about pirates in which a young woman is kidnapped from the ship on which she is traveling. Rescued by her hero, both are shipwrecked on an island inhabited by dinosaurs (filmed material taken from the stock of *One Million B.C.*, 1940). In *Unknown Island*, an expedition arrives on a South Pacific island as a result of evidence from aerial photographs that appear to indicate the existence of dinosaurs. The island's dinofauna includes a dimetrodon (not a dinosaur), a tyrannosaur, and a ferocious herd of ceratosaurs.

Popular literature from the beginning of the 20th century includes many stories of lost worlds. Doc Savage is a well-known superhero of the pulp novels written by Kenneth Robeson in the 1930s. His adven-

Still from the film *King Kong* (RKO Pictures, 1933). The enormous gorilla is the indisputable king of a lost world populated by dinosaurs and other beasts from our remote past.

ture, entitled *The Land of Terror* (1933), involves the struggle of Savage and his formidable friends against a gang of delinquents, whose chief has discovered a substance that is capable of dissolving anything—from a person to the steel of a strongbox—in a matter of seconds. One of the main components of this substance is only found in a crater on a solitary island in New Zealand, where prehistoric flora and fauna have been preserved. Savage's erudite explanation of the phenomenon is that the territory formed part of the Asian continent and life entered the crater through a place of access that disappeared over time; the current island appeared as a result of Earth sinking and the oceans crashing in. As well as dinosaurs and other prehistoric beasts, the interior of the crater is home to an exuberant flora similar to that of the Carboniferous, maintained thanks to the misty atmosphere produced by volcanic activity. In the film *The Lost Continent* (1951), an expeditionary group in search of a lost missile comes across a great plain inhabited by dinosaurs. In this case, intense radioactivity explains the survival of the prehistoric fauna.

One of the writers to have made the greatest contribution to the generation of the myth of the lost world was Edgar Rice Burroughs (1875–1950), the famous U.S. creator of Tarzan and the inventor of the lost world of Pellucidar. Pellucidar is a prehistoric place located at the center of the Earth, with an interior sun that illuminates a concave world. The forests, the mountains, and the oceans of this world are inhabited by extinct animals from all the eras of our geologic past: dinosaurs, pterosaurs, mammoths, saber-toothed tigers, and various types of primitive humans. In the world of Pellucidar, there is also a group of evil Spanish pirates, headed by a ringleader named *El Cid*, who were undoubtedly visitors who had been detained there. The Pellucidar series, made up of seven novels, began in 1922 with *At the Earth's Core*, a screen version of which appeared in 1976. In this first episode, the adventurer David Innes and the geologist Perry Abner arrive in Pellucidar on board a vehicle that bores into the Earth. This means of gaining access to a lost world has been used on other occasions, such as in the film *The Last Dinosaur* (1977). Pellucidar is also the setting for one of the adventures of a Tarzan novel (*Tarzan at the Earth's Core*, 1930). It is not the only lost prehistoric world that Tarzan visited. In the 1920 novel *Tarzan the Terrible*, the famous ape-man goes deep into the land of Pal-ul-don in search of his companion, who has been captured by German troops. The land of Pal-ul-don is inhabited by various races of primitive humans and the *gryfs*, modified dinosaurs of the genus *Triceratops*. Finally, the inventive Burroughs developed a third literary lost world, the land of Caprona, in the course of three novels published in 1918: *The Land That Time Forgot* (a film of which appeared in 1974), *The People That Time Forgot* (film version 1977), and *Out of Time's Abyss*. Caprona (also known as Caspak) is a continent lost in the South Seas (darkly referred to by an Italian sailor from the 18th century), inhabited by enormous ferns and insects, plesiosaurs, pterosaurs, various kinds of dinosaurs, long-haired rhinoceroses, cave hyenas, aurochs, and various hominid species. It seems evident that this mixed prehistoric fauna is one of the constants of Burroughs's lost worlds, although it is a feature present in many other stories. In the case of Caprona, the explanation is founded on a metaphorical comparison between time and a river. As one heads north, upstream of Caprona's river, one comes across increasingly evolved creatures, culminating in *Homo sapiens*.

The film *The Land That Time Forgot* locates the lost world in the frozen waters of the South Seas. *The Land Unknown* (1957) does the

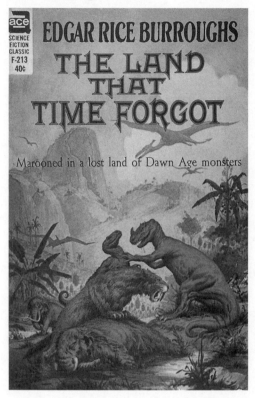

The cover art for the paperback *The Land That Time Forgot* by Edgar Rice Burroughs.

same, although in Antarctic waters. A helicopter carrying four people crashes in a prehistoric valley after colliding with a pterosaur. In this place, situated 1,000 m below the edge of the ice mass, the survivors have to defend themselves against a tyrannosaur and an enormous plesiosaur (an elasmosaur, according to various film commentators).

One of the most obvious properties of a lost world should be the slow rate of evolutionary change. This means that if organisms have stayed alive for tens or hundreds of millions of years, they must scarcely have changed over time. Doc Savage explains this phenomenon in *The Land of Terror*. The famous adventurer's conclusion is radically environmentalist: if an animal lives in a warm country, its hide will be thin and light, or it may not even have a hide at all. However, if the country turns cold, it must develop a thick hide or die. This acquisition is the result of evolution, Savage states. But there have been no changes within the crater: the environment remains the same as ever, and the trapped organisms "have not needed to evolve." Nevertheless, in other cases,

The Land Unknown (Universal International, 1957). This film had one of the worst "human-in-a-costume" impersonations of *Tyrannosaurus*.

the animals of the lost world have clearly become transformed in relation to their ancestors. In the land of Pal-ul-don, the *Triceratops* have turned into carnivores, the original hooves of their legs having become powerful claws. In Pellucidar, the *mahars*, an intelligent telepathic race, are descendants of the pterosaurs.

Another odd trait of the lost world is its sense of being a prohibited and damned place, dangerous for those who dare to venture into it and the source of dramatic events if one of its inhabitants abandons the place (in this context, remember the impressive opening of the door in the wall that separated King Kong from the rest of the world). The curse of the lost world stems from its ominous character. It is a place in which humans are not the dominant species and in which humankind is scarcely able to survive against the enormous beasts of the past. The myth of the lost world is also in some way linked with that of Paradise after the departure of Adam and Eve. The lost world is a morally and physically dangerous place. Dinosaurs have on occasion represented primordial evil—the fundamental evil that has existed since the beginning of time.

The clearest example of this is the allosaur Gwangi from the film *The Valley of Gwangi* (1969). This valley, located in México (although really it is in the "Ciudad Encantada," the Enchanted City, of Cuenca, Spain), features a small ancestor of the horse (*Eohippus*), large carnivores, ornithomimids, styracosaurs, and enormous pterosaurs of the genus *Pteranodon*. The action takes place in 1912. The story describes how a group of cowboys manage to trap the enormous carnivorous dinosaur and put it on show in a bullring, although not without having received a warning from an aged gypsy woman about the curse that hangs upon the dinosaur. Finally, Gwangi manages to escape, and in pursuit of the crowd, he ends up entering a cathedral (in reality, that of Cuenca). He is destroyed by the flames within the sacred place, an obvious symbol of a purification process that puts an end to the forces of evil. As on other occasions, *The Valley of Gwangi* does not offer any explanation about how the dinosaurs have survived. Their preservation in a frozen or suspended state is an entirely intuitive idea.

11

Frozen Dinosaurs

It is likely that the various discoveries of mammoths in the Siberian permafrost were one of the factors responsible for the process of synchronization between humans and dinosaurs within the fantasy discourse. All that is required is to swap the proboscidians (mammoths, elephants, mastodons, etc.) for dinosaurs. This is the approach taken in the story by Spanish science fiction writer Alan Comet (the nom de plume of Enrique Sánchez Pascual) in his 1955 novel *El Despertar del Pasado* (*The Waking of the Past*). It is the 21st century and the Soviet Union has been defeated, leaving a few communists hiding out in an underground base in Siberia. Professor Komarow, who revives human beings by inserting a small electrode into their brains, manages to bring back to life thousands of "monsters of the past" that were sleeping in the ice. The communists set this terrifying horde loose on Europe in order to defeat their enemies. In the end, the United States air force puts a stop to it, and it also manages to return the "monsters" to Siberia and put an end to the Soviet peril once and for all.

The existence of a frozen dinosaur immediately spells potential danger for humanity. This is the sense of films such as *The Beast from 20,000 Fathoms* (1953) and *Reptilicus* (1962). In the former, a nuclear explosion in the Arctic awakens an enormous dinosaur that has been locked in the polar ice for 100 million years. A scientist sees the animal, but no one believes him. Finally, the scientist speaks to a Canadian sailor whose ship has been attacked by a strange animal. The animal (a "marine dinosaur") heads south along the Atlantic seaboard of the United States and Canada, destroying a lighthouse in Nova Scotia. Meanwhile, the testimonies of the scientist and sailor convince a famous paleontologist and the army that the monster really does exist.

The Beast from 20,000 Fathoms (Warner Brothers, 1953). The beastly nature of the dinosaur (see chapter 21) is clearly stressed in the film poster. The alteration of the social and natural orders—"It's alive!" says a terrorized character—is depicted too (see chapter 20).

After identifying the type of dinosaur (a *"Rhedosaurus"*), the paleontologist reveals that it is a relative of the sauropods and that remains of this type of dinosaur had been found in a submarine canyon on the New York coast. The paleontologist locates the dinosaur with a bathyscaphe, but the dinosaur then eats the scientist. (This is the first and only time that a dinosaurologist has been used as bait to catch a dinosaur.) Finally, the *"Rhedosaurus"* appears in the New York docks, where it eats a police officer and sows panic among the population. After the animal is injured at the base of its neck by a bazooka, the health authorities discover that its blood carries a serious infection that can affect human beings. This problem leads the scientists to kill the monster by introducing a radioactive material–laden projectile into its wound. An enormous fire finally does away with the dinosaur on the Coney Island fun fair's roller coaster. *The Beast from 20,000 Fathoms* is based on the short story "The Foghorn" (1951) by Ray Bradbury, and it achieved well-deserved commercial success; it became a cult movie and an obligatory point of reference in the history of science fiction cinema. In Bradbury's short story, an enormous dinosaur that has emerged from the deeps visits a lighthouse every year. Its foghorn faithfully reproduces the mating call of the species to which the dinosaur belongs.

In the film *Reptilicus*, some miners from Lapland discover blood and fragments of skin and bone from a large animal in a borehole. The material is sent to an aquarium in Copenhagen. The entire animal is regenerated from the tail in an enormous tank. It is a snakelike dinosaur, 27 m long, that will finally be destroyed only after spreading panic throughout Copenhagen.

At times, the object that allows a dinosaur to be brought back to life is not the adult organism itself but an egg. This is the case in the story of *The Dinosaur and Love*, by R. Arkham, in which an egg discovered in an Alaskan snowdrift is deposited in a museum. The state of suspended animation implies, in fact, a journey through time at *normal* speed. In some cases, the scientists in fantasy stories have been able to travel through time at high velocity, on many occasions encountering organisms from our remote past.

12

Time Travels

Time travel is one of science fiction's great themes. When the leap back in time is considerable, it seems the human beings in the fantasy always come face to face with dinosaurs, whatever the geologic era. The encounter with the beasts of the past may happen intentionally, as in the two stories I discuss below, or it may be an unintended consequence of scientific research. Ray Bradbury's short story "A Sound of Thunder" (1952) takes place in the more or less near future. A safari company organizes hunting trips to the past: the client chooses his animal, they take him back in time to when the animal was alive, and the hunter shoots the animal. One such hunter, fleeing in terror in the face of an imposing tyrannosaur, treads on a butterfly. This tiny accident, because of the multiplicative effect of unforeseeable circumstances, results in an important alteration in the present time when the chrononauts return—the language is spoken differently and a different President of the United States has been elected.

Equally, dinosaurs have been instruments of political power, as happens in Malcolm Hulke's story *Doctor Who and the Dinosaur Invasion* (1978), a novelization of the *Doctor Who* British television program entitled "Invasion of the Dinosaurs" (first broadcast in 1974), which Hulke also wrote. A group of great humans, discontented with the civilization of their times, which they consider to be dehumanizing and against Nature, decide to build a machine beneath the center of London—the Timescoop—that is able to turn back time. With this machine, they intend to return to a supposed golden age of humanity, which they would direct thereafter in order to avoid the age-old mistakes. In order to ensure that only people they select end up making this voyage, they forge a plan to cause the majority of the population to abandon the center of London. This plan consists of making a wide variety of dinosaurs appear periodically in the center of the city, which the army has to seek out and destroy. Fortunately, these evil plans, in which the rest of humanity not controlled by the Timescoop are to disappear, are ruined by the intervention of the Doctor.

In some cases, scientific research into the space-time continuum may have consequences that are not closely related to this field, as occurs in Isaac Asimov's short story "A Statue for Daddy" (1958). In this tale, a temporal connection with the Mesozoic (the "chronotube") allows two scientists to recover 14 dinosaur eggs. The eggs are carefully incubated, and they hatch bipedal dinosaurs about the same size as a medium-sized dog. One of them is accidentally electrocuted, and the scientists discover that their meat is truly exquisite. They become fabulously rich rearing and marketing dinosaur meat under the name of "dinochicken."

In Arthur C. Clarke's story "Time's Arrow" (1952), a group of paleontologists unearth the ichnites (footprints) of a large carnivorous dinosaur. A few kilometers from the place a physics laboratory is investigating time travel with "Helio II," a substance possessing negative entropy. The head of the paleontologists is at the laboratory, his off-road vehicle parked outside, when there is an enormous explosion. At the excavation site, the paleontologists have just brought the last footprints of the great theropod dinosaur to the surface. The animal rapidly changes the route and pace of its movements: some dinosaur tracks are found overlaying the paleontologists' off-road vehicle's tire tracks.

In the film *My Science Project* (1985), an enormous carnivorous dinosaur travels through time, thanks to the power of a mysterious extraterrestrial engine.

In all of the stories and films mentioned, temporal overlap is achieved by means of an apparatus that is conventionally known as a time machine. Nevertheless, some chrononauts use other methods — as, for example, the main character in Steve Utley's short story, "Getting Away" (1976). Here, the procedure used is based on the appearance of a disease known as "chronopathy." Retrospective chronopathic visions lead the protagonist to time-travel and to lodge himself within the body of a person or an animal. In this way, the narrator of the story sees trilobites through the eyes of an armor-plated fish from the Devonian; he flies inside a pterosaurs over the Cretaceous sea of Kansas; and he feels that he rules the world as a 20-m-long dinosaur. The main character lives in a highly polluted future. It is a world that is about to die out because the self-destructiveness of human beings means the end of life on Earth, in contrast to the dinosaurs, which disappeared with dignity and left behind a clean world.

In his novel _The Dechronization of Sam Magruder_ (1996), George Gaylord Simpson describes the process by which the lead character is accidentally transported back to the Cretaceous. Magruder believes that time has a discrete nature, rather than being the continuous flow perceived by humanity. The chronologist manages to slow down the passage of time until the discontinuous units of time become perceptible. Magruder has the misfortune to slip between two discrete units of time and finally to journey toward the end of the Mesozoic. There, he has the opportunity to hunt, watch, and admire the varied forms of dinosaur, especially ceratopsians of the genus _Pentaceratops_ and ornithomimosaurs (_Struthiomimus_).

Another means of traveling to our remote past has been by allegory. In the Czech film _Journey to the Beginning of Time_ (1955), four boys enter a cavern that leads them to a river—an allegory of time. Their intention is to reach the Silurian, 400 million years ago, to see living trilobites. After crossing typical Quaternary and Tertiary landscapes, they arrive at a spot on the river with plants and animals from the Mesozoic. They are then attacked by a group of pteranodons and observe various dinosaurs, including a carnivore and a stegosaur fighting each other. The film has a strong didactic sense throughout and faithfully adheres to the scientific postulates of the time (e.g., aquatic sauropods and hadrosaurs, dinosaur stupidity). Even so, the story brims with a special affection and feeling for paleontology that would move any lover of the science of fossils.

Sometimes, interactions between humans and dinosaurs are not possible, but the scientists engineer them to be able to observe the appearance of the animals. The novel _Before the Dawn_ (1934) by the Scottish mathematician John Taine (whose real name is E. Temple Bell) is a classic example of the capacity for capturing images from the past. The basic idea is that the light has recorded scenes on ancient materials that in the present day can be viewed with an apparatus that reads this evidence and projects it in three dimensions. The real star of the story is a tyrannosaur that the spectators of the past name Belshazzar. The story of the dinosaur is described in almost epic terms and represents the heroism of a species fighting the threat of its extinction.

In the interesting Spanish film _Horror Express_ (1972), set in 1907, Professor A. Saxton (Christopher Lee) discovers a 200-million-year-old frozen hominid in China. In fact, the fossil houses an alien that arrived on Earth during the Mesozoic and is able to pass from one body to

Journey to the Beginning of Time, a Czechoslovakian film by Karen Zeman. There were two versions, the original and an American version with extra footage shot at the American Museum of Natural History in New York.

another. Saxton and Wells (Peter Cushing) look at the hominid–extraterrestrial's ocular fluids under the microscope and are amazed to see the great animals of the past (which are really old drawings of dinosaur reconstructions). (We should remember that the hypothesis that the final image seen by someone before they die remains recorded on their retina was put forward by Dr. Bourion at the French Society of Legal Medicine in 1865.)

Any chrononauts who want to see dinosaurs in their own world must move to the period of time between the Upper Triassic (some 230 million years ago) and the Upper Cretaceous (around 65 million years ago). Most dinosaurs, as the reader well knows, disappeared during the latter period, an era characterized by a great biotic crisis, but science fiction has advanced the possibility that some dinosaurs (other than the birds, of course) could reappear in the future.

13

Dinosaurs of the Future

The English naturalist Charles Lyell (1797–1875) is considered to be the founder of modern geology. His work *Principles of Geology* (1830–1833) had an enormous influence in a wide variety of fields, as well as on the thoughts of Charles Darwin. Lyell was involved in what came to be called the "progressionist" debate, or in other words, the conception of historic organic change as a process with a direction—progress—leading from the simplest to the most complex animals, with humanity at the summit. Lyell's stance was antidirectionist: he believed that the history of organisms and the physical world developed cyclically. In this way, it was possible to believe that one could discover trilobites from the Tertiary, even though they became extinct in the Paleozoic, and even that dinosaurs would reappear in the future. This basic idea, a hypothesis that has now been discarded, forms the starting point for various fantasy stories, even though they go beyond the context of Lyell's cyclicity.

At the beginning of the 20th century, the Franco-Belgian paleontologist Louis Dollo (1857–1931) proposed his "Law of Irreversibility" in evolution. According to this principle, if a particular structure of an organism atrophies, it will never develop again, and if it disappears, it will never reappear in the organism's descendants. In general terms, this rule of evolutionary biology implies that it is impossible for the complex genomic information responsible for the existence of a group of living organisms to reappear once they have disappeared. This principle assumes the impossibility of the reappearance of nonavian dinosaurs in nature. If Lyell's cyclical model cannot be invoked to explain the dinosaurs of the future, what can be? There are two possible answers: catastrophic natural causes, and disasters generated by all-out nuclear wars. Within the first category is included J. G. Ballard's novel

The Drowned World (1962), a story with a prophetic beginning: the gradual degradation of the layers of the ionosphere in the 21st century produces a terrible increase in solar radiation. With the increase in temperature and the melting of the poles, Europe becomes a land of tropical swamps. In this new environment, a flora and fauna equivalent to those of Triassic times is reborn.

This is the cover art for the classic pulp magazine *Famous Monsters of Filmland* (Warren Publishing Co.). It features the Japanese film monsters the Bionic Monster and Godzilla (Toho).

In Roger Corman's film *Teenage Caveman* (1958), the actor Robert Vaughn portrays a boy from a tribe of primitive human beings who journey into a prohibited area, where they come up against monstrous animals. At the end of the film, it is revealed that the story is not set in the remote past but in the future, after a nuclear war has returned humanity to a primitive state. The films *Yor the Hunter from the Future* (1982) and *A Nymphoid Barbarian in Dinosaur Hell* (1993) also take place in such a postapocalyptic environment. In the latter film, the dinosaurs of the future arise through terrible mutations that turn inoffensive mammals into gigantic beasts.

The improbability of dinosaurs reappearing in our world in the future has already been mentioned—but could they appear on another planet?

14

Exodinosaurs

To contemplate the possibility of the existence of dinosaurs on another planet (exodinosaurs), it is necessary to postulate that under similar conditions, the processes of organic evolution give rise to similar beings. This postulate means that on a planet that has a similar historical development and similar environments to that of Earth, the living organisms that arise over time will be similar. The idea underpinning this concept is essentially adaptationist. In a given environment, only the best-adapted forms will survive. Because the problem of what the best adaptation is has a unique solution (in terms of the morphology, physiology, and behavior of living beings), the result must be similar organisms. This is the position defended by the Russian paleontologist and science fiction author Ivan Antonovitch Efremov (1907–1972). In his novel *Star Ships* (1948), extraterrestrial visitors encounter dinosaurs in the Upper Cretaceous. These visitors are similar to human beings, the "supreme form of thinking material." Throughout the entire universe, living organisms endowed with self-awareness have to have an appearance similar to our own.

Without a doubt, taking these assumptions into account, it is far from strange that human travelers encounter dinosaurs on other planets, as in the films *King Dinosaur* (1955) and *Planet of Dinosaurs* (1978). In the first, four astronauts land on the planet Nova, where they confront a host of prehistoric beasts, among which is an enormous dinosaur, played by an iguana enlarged to giant proportions by cinematic techniques. Horrified by the savage appearance of the planet, the astronauts decide to destroy it with an atomic explosion. The initial approach of *Planet of Dinosaurs* is similar, although the denouement of the story is much more rational. A group of astronauts manages to land on an unknown planet, similar to Earth. They splash down in a large body of

The cover of the December 1940 issue of *Astonishing Stories* featuring the fictional dinosaur *"Centosaurus."* The story was by science fiction Grand Master Isaac Asimov.

King Dinosaur (Lippert, 1955). This film was so comical that it was spoofed by the television series *Mystery Science Theater 3000* (Best Brains Productions).

water and have to get out of their ship with great haste, taking with them only the most basic of survival kits. After various encounters with all manner of dinosaurs (including brontosaurs, stegosaurs, centrosaurs, and ornithomimosaurs), they decide that the only way they can survive is to challenge a large theropod that is the dominant predator in the land in which they find themselves. They finally manage to kill it by using themselves as bait and attracting the huge dinosaur to a poisoned stake. The final sequences of the film show the group of surviving humans (two women and three men) ready to start a new life.

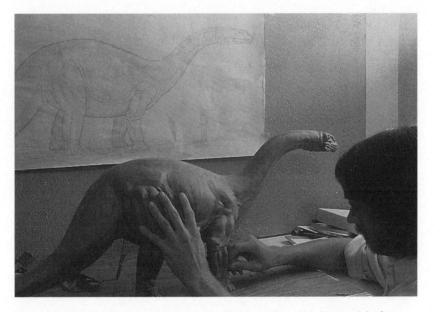

The U.S. dinosaurologist and cineaste Stephen Czerkas with his model of a sauropod for the film *Planet of Dinosaurs* (1978). Courtesy of Stephen Czerkas.

The action of the film *The Planet of the Storms* (1962) takes place in 1985. Humankind has conquered the Moon and Mars, and now two Soviet spacecraft are heading toward Venus. There, they are attacked by giant plants and dinosaurs: the planet is going through a geologic period similar to the Earth's Mesozoic.

The films mentioned above assume that the exodinosaurs are the same as those that existed on Earth. Big monsters that could be interpreted as being dinosaurs belonging to other planets turn up in other films. A succinct list would include *Twenty Million Miles to Earth* (1957), *The Return of the Jedi* (1983), and *Coneheads* (1993).

The advanced level of scientific knowledge that allows a person to be put on another planet would also make it possible for dinosaurs to reappear, reclaiming their genome.

The New Frontiers of Science: *Jurassic Park*

In an article that appeared in the weekly magazine *Newsweek* (June 1993), journalist Sharon Begley suggested that the best science fiction is based on the dominant scientific paradigm of its time: the creature invented by Mary Wollstonecraft Shelley, Frankenstein's monster, is brought back to life by means of electricity, and Godzilla, the enormous Japanese dinosaur-dragon, is revived by the radioactivity of an atomic bomb. Thus, the dinosaurs of Michael Crichton's 1990 novel *Jurassic Park* are generated through the medium of biotechnology.

The fossil record (information about life in the past that has been retained in sedimentary rocks) contains a host of "documents," including bones, teeth, shells, footprints, and eggs. Paleontologists discovered many years ago that even organic matter can be found in the fossil record and that under certain conditions, it can survive with few or even no modifications for long periods of time. In this way, two proteins (osteocalcin and collagen) have been found in dinosaur bones that are associated with tissues that make up the structures that supported the animal. Could we even find the genetic material (the DNA) of a dinosaur? The first place it would occur to anyone to look would be in a bone. Unfortunately, nobody has so far been able to prove that the DNA found in dinosaur bones is not due to external contamination—for example, by bacteria or fungi. The nucleus of the idea of *Jurassic Park* is that dinosaur DNA is obtained from individuals of blood-sucking insects that fed on these animals and were then preserved in amber, a fossil resin that may be more than 200 million years old. The DNA extracted in this way is degraded and could not represent more than part of the original. So the scientists in *Jurassic Park* fill in the regions lacking information with genetic material derived from present-

Godzilla, King of the Monsters (Toho, 1954). There were two versions of this film, the original, and an American version with extra footage added featuring actor Raymond Burr.

day reptiles and amphibians. Once reconstructed, the dinosaur DNA is implanted into the egg of a crocodile that has been stripped of its own genetic material. They even build an artificial plastic shell for this egg. Having taken the appropriate scientific steps, the birth of a dinosaur is assured.

The methods Michael Crichton describe in his novel are all fiction—and highly ingenious fiction at that. Unfortunately, scientists have not yet reached the stage where the process described above could be carried out. Parasitic insects have not yet been found containing dinosaur blood, although some researchers claim that they have identified some species of Diptera (chironomids) that are contemporaries of the dinosaurs and are adapted to parasitizing large terrestrial vertebrates. Nor do we currently know how to fill in the numerous absent fragments of fossil DNA by using that of present-day organisms. Finally, even if complete dinosaur DNA were available, current science would not be capable of implanting it in a foreign cell as a procedure for cloning a dinosaur. Some scientists and the director of the 1993 *Juras-*

This is the first full view of a computer-generated dinosaur (*Brachiosaurus*) in the ground-breaking film *Jurassic Park* (Universal Studios, 1993).

sic Park film version, Steven Spielberg, are convinced that these current difficulties will be overcome in the near future, in a period of 30 to 40 years at the most, and indeed, this may possibly turn out to be so. However, in the more immediate future, dinosaur lovers can enjoy the magnificent reconstructions of these animals that Spielberg has created for the screen.

16

The Extinction of the Dinosaurs

The concept of extinction—that is to say, the disappearance of entire lineages of living organisms—is intimately linked with the study of the fossil record. In effect, it was the great French naturalist Georges Cuvier who, at the beginning of the 19th century, showed that forms such as the mammoth and other large terrestrial vertebrates had become extinct. The first paleontologists to describe dinosaur bones knew that the living organisms from which the fossil bones came had disappeared from the planet, and paleontology has since spent more than a century trying to establish why. What caused the extinction of the dinosaurs? This question, which continues to be a favorite of the media, still has no clear answer. The disappearance of the majority of the dinosaurs happened at the end of the Cretaceous, some 65 million years ago. A huge number of other organisms went extinct along with them: large reptiles (pterosaurs, plesiosaurs, mosasaurs), ammonites (cephalopods with external shells), rudists (reef-building bivalves), and most of the foraminiferans (single-celled marine organisms with a complicated shell). In short, it has been calculated that approximately 50% of the species existing at the end of the Mesozoic era became extinct.

Many hypotheses have been advanced to explain this great biotic crisis, and that of the dinosaurs in particular. Of the hundred or more suggestions put forward in the last few decades, the most widely accepted fall into two categories of causes: internal and external. The former refers to the supposed problems that arose within the dinosaurs themselves, their structure and the process of their evolution, and are closely connected to the orthogenetic hypotheses mentioned previously: for example, that the intervertebral disks of the dinosaurs were not well adapted, or that the dinosaurs became too big or too strange and thus became extinct. Naturally, there is no evidence to confirm any of these proposals.

The external hypotheses concern particular environmental factors that could have done away with the dinosaurs. For example, it has been repeatedly suggested that the mammals could have brought about the end of the dinosaurs because they ate their eggs. Of course, there is no evidence to support this hypothesis. We also have to bear in mind that the first mammals appeared at the same time as the dinosaurs, about 230 million years ago. Both groups coexisted for 160 million years. If this trophic dependence really existed, then why did it only manifest itself at the end of the Cretaceous? The group of external hypotheses encompasses the viewpoints that are currently most widely accepted as being most likely: gradual climatic changes and the fall of an extraterrestrial object—a meteorite. The first concerns the persistence of a less extreme climate during the Upper Cretaceous than that of the present day, without such seasonal differences. It is thought that the existence of shallow but extensive epicontinental seas would account for a climate milder than our current one. However, at the end of the Cretaceous, the sea retreated considerably, abandoning the epicontinental basins, or internal seas. This phenomenon, which could have lasted for around 100,000 years, implies that a more extreme climate arose, with colder winters and warmer summers. Dinosaurs are believed to have been unable to withstand the environmental changes and so disappeared. Some questions immediately come to mind, such as why did lizards, snakes, crocodiles, and tortoises not disappear as well? And why did the winged dinosaurs with feathers (birds) not also become extinct? Neither question can be answered satisfactorily by applying either environmental explanations or, as we shall see, that of the collision of a meteorite.

In recent years, it has been shown that the Cretaceous-Tertiary boundary sediments contain an abnormally high level of iridium, an element from the platinum group that is found in the deep layers of the planet. Consequently, scientists put forward the hypothesis of the collision of a large extraterrestrial body, reasoning that the iridium must have come from outer space. The main variables that have to be considered in respect of such an impact are the size, density, and velocity of the meteorite and the density of the substrate receiving the impact. An object 10 km in diameter with a density of 2,200 kg/m^3 and a velocity of 25 km/s would produce a crater some 100 km in diameter in basaltic rocks and 122 km in sedimentary rocks. The impact would liberate a quantity of energy around 1,000 times that of the entire human nuclear arsenal and would produce an enormous dust cloud around the planet. The iridium would have come from this cloud, which would have prevented plants receiving sunlight during a blackout that could have lasted for up to a year.

Naturally, the disruption in the plant world would have immediately affected animals. The dust cloud would also have caused a marked cooling (a "nuclear winter") that would have had additional lethal effects from the outset. This is the frightening scenario of the meteorite hypothesis, which was proposed at the beginning of the 1980s. Since then, many scientists have looked for real evidence of this hypothesis. The new techniques of geophysics and teledetection have allowed the confirmation of the existence of numerous craters of enormous size (20 to 200 km) distributed over all continents. In the 1990s, a great deal of work was done in a study of the Chicxulub crater (Yucatán, México), which seems to fulfill all the requisites for being recognized as being due to an event that could have caused the extinction of the dinosaurs. The size of the Yucatán crater could be between 170 and 300 km, according to which opinion you go by, and the time of the collision has been dated at 65 million years ago. Evidence of tsunamites — deposits formed by giant waves — has been found in Texas, Haiti, and other places in the Caribbean basin. These sediments originated from the impact of a meteorite in what was originally the sea (the "proto-Caribbean").

Is there any evidence that would allow us to decide between the two hypotheses? Climate change implies a gradual extinction over hundreds of thousands, or even millions, of years. The fall of an extraterrestrial object is a rapid event and would have driven the dinosaurs and other organisms to extinction over a short period, of months or even less. Therefore, we can look for confirmation of this rate of disappearance in the fossil record. Unfortunately, it seems that paleontologists cannot reach agreement on this point. There are opposing opinions with respect to the foraminiferans and the dinosaurs. The fact that both are based on observations of the fossil record suggests that more work needs to be done. So we still do not know if dinosaurs underwent a slow decline in their diversity, an agony that could have lasted from 10 to 15 million years until their extinction, or if an extraterrestrial agent erased them from the planet.

Extraterrestrial agents of a nature other than meteorites or comets are one of the essential reasons put forward to explain the extinction of the dinosaurs in the genres of fantastic realism and science fiction. Fantastic realism is a heterodox interpretation of the history of humankind and culture that enjoyed a certain standing in the 1960s and 1970s.

One of its essential premises is the belief in the historical interaction of extraterrestrials with human beings, and even their former presence on our planet. Some authors, such as E. von Daniken and J. J. Benítez, have described the presence of human footprints next to dinosaur tracks in the United States, New Zealand, and Spain. These pseudoscientific ideas have recently been revived by the publication of the book *Darwin's Error* (1998) by the German fundamentalist engineer Hans-Joachim Zillmer, which contains ridiculous interpretations of the fossil and lithologic records. Thus, the belief in the synchrony of humans and dinosaurs belongs not only to fantastic realism, but also to particular viewpoints of religious fundamentalism.

Creationism is a political instrument of religious fundamentalism that seeks to impose a body of religious beliefs derived from a literal interpretation of the Bible. One of its principles is the denial of any evolutionary dimension to life, the Earth, or the universe. At the beginning of the 1970s, the fundamentalists tried to reinforce their strategy through the development of a religious pseudoscience known as creation science. This pseudoscience has given rise to books and videos of its peculiar interpretation of the fossil record—and in particular of dinosaurs. The organization Films for Christ has released *The Great Dinosaur Mystery* and *The Fossil Record* (both 1995). The first of these is a milder version of P. S. Taylor's book, *The Great Dinosaur Mystery and the Bible* (1987). Creationist dinosaurology holds that the bodies of the first dinosaurs, like that of Adam, were directly created by God from dust and mud. Therefore, the dinosaurs also occupied their place in the earthly paradise alongside the first pair of human beings. The dinosaurs of paradise were defenseless and peaceful creatures that ate plants and fruit. Like other animals, their reason for being was to promote the happiness of the human beings. Other creationist ideas defend the view that the reference in the Bible to the creature known as Behemoth, which God shows to Job, is a description of a dinosaur. This biblical character lived after the flood, which means that Noah must also have saved the dinosaurs. To get around the problems arising from the dinosaurs' immense size, Noah opted to include young specimens, of reduced size, in his ark.

In *Les Extra-Terrestres dans l'Histoire* (*Extraterrestrials in History*) (1972), J. Bergier claimed that dinosaurs were extinguished by intelligences from beyond this planet through the generation of a nova, the explosive birth of a star. Surprisingly, the reasons put forward by Bergier clearly belong to the realm of the ideas of transcendental finalism. In effect, the famous French author of fantastic realism affirmed that evolution on Earth had followed pathways that ended up in a cul-de-sac because there had not been any progress in the large reptiles over 150 million years. The extraterrestrials, wanting to increase the number of intelligent species in the universe, destroyed the dinosaurs to allow humankind to emerge. From the moment that Bergier accepted that these alien beings could be called gods, his outlook became similar to the ideas of paleontologists such as P. Teilhard de Chardin and M. Crusafont. The Crusafontist concept of background orthogenesis is that of an evolutionary line that, from the beginnings of life, passes through particular types of fish, amphibians, and reptiles, and leads directly to humans. Dinosaurs were far off this line and were therefore condemned to extinction.

In various science fiction stories, individual dinosaurs are destroyed and the species even totally wiped out by beings from other planets. In Clifford D. Simak's novel *Our Children's Children* (1974), humanity of the 25th century moves to our time through time tunnels. Their reason for doing so is that they are fleeing from an invasion of extraterrestrials against whom they cannot defend themselves: they are ferocious aliens endowed with a murderous mysticism. In the end, the threat disappears, but the extraterrestrials travel to the Upper Cretaceous, where they cause the extinction of the dinosaurs. Ivan Antonovitch Efremov's tale *Star Ships* (1948), briefly discussed in chapter 14, suggests that the star systems are enormous craft that transport through the universe the inhabitants of the planets that accompany the stars. In search of sources of energy, alien beings arrive on Earth at the end of the Cretaceous. This period is characterized by the appearance of the great mountain ranges that we know of today as the Himalayas, the Caucasus, the Alps, etc. According to the paleontologists in Efremov's novel, the orogenesis was produced by radioactive decay of superheavy elements in the Earth's crust. The energy of these atomic reactions could have been liberated at the surface of the planet. These are the sources of energy sought by the extraterrestrials, who also become responsible for the extinction of the dinosaurs. The arrival of the aliens results in a serious conflict with the dinosaurs, and the aliens decimated them. The proofs of this conflict are evident: enormous cemeteries discovered by Soviet paleontologists in eastern Russia and central Asia, and bones perforated by wounds caused by projectiles. Finally, the scientists find the cranium of a humanoid and some advanced tools next to the skeletons of the dinosaurs.

In particular cases, the relationship between aliens and dinosaurs is pure scientific curiosity. In A. Porges's story "The Ruum" (1955), an extraterrestrial spaceship abandons an artifact on Earth in the era of the dinosaurs. In our time, a uranium prospector discovers an enormous concentration of animals in suspended animation, among which are organisms that have existed during the last 80 million years. It turns out to be the work of an extraterrestrial artifact (the *ruum*) that, on finding the lone miner, sets out in pursuit of him. In the end, it catches up with him but does nothing to him. The miner has not found any uranium but will become rich by selling dinosaurs to scientific institutions.

The humanity of our current times is responsible for the disappearance of the dinosaurs in Geoffrey A. Landis's tale, "Dinosaurs" (1985). A secret office of the U.S. federal government is devoted to analyzing the possibilities of using people with paranormal powers. Among them there is a child who can predict the immediate future and who is able to make objects disappear, in such a way that they are transferred to an unknown place or time. The child prevents a Soviet nuclear attack by making it disappear from our time, moving 6,000 nuclear warheads to the end of the Cretaceous. We should note that the consequences produced by these weapons would be similar to those previously mentioned of the collision of a meteorite.

In some of the ideas proposed in science fiction, the threat to the dinosaurs comes neither from aliens nor humans, but from the dinosaurs themselves. Isaac Asimov's story "Day of the Hunters" (1950) describes how the dinosaurs were driven to extinction by a lineage of intelligent reptiles capable of inventing powerful armaments with which they hunted for pleasure. This idea has much to do with the unlikely reflection offered by Tarzan in Edgar Rice Burroughs's novel *Tarzan at the Earth's Core* (1930), in which carnivorous dinosaurs do away with the herbivores and then eat each other up.

However, are all the nonavian dinosaurs extinct? Ray Bradbury, in "The Foghorn" (1951), maintains that this is not the case and that the dinosaurs took refuge on the ocean beds. This idea is connected with the suggestion that lost worlds, or at least part of their fauna, exist—not in literature or cinema, but in reality.

17

Nessie and Friends

On December 23, 1919, a retired officer of His Britannic Majesty's army, Captain Leicester B. Stevens, took the train from London's Waterloo Station to Southampton. From the coast, he boarded a ship to Cape Town, from where he headed north to the Congo, which in those times and for many years after was a Belgian colony. His aim was none other than to hunt a dinosaur. His arms comprised various high-caliber rifles and a ferocious animal, a cross between a German shepherd dog and a wolf, trained by the Kaiser's army during the Great War, which had ended the previous year. Captain Stevens also took along several butterfly nets. This latter fact was revealed by the London newspaper, the *Evening News*, in an article of December 12 about a man armed with a butterfly net setting out to hunt the thunder lizard. Captain Stevens naturally knew that a butterfly net was not suitable for catching a large-sized dinosaur, but he probably reasoned that if he did not find one, then he could spend his time collecting insects. In fact, the English officer mysteriously declared that he knew the enormous reptile's vulnerable spot, at which he should shoot. The whole story provoked an authentic flood of letters and comment in the British press. A general advised Stevens to take a small field gun. An elderly lady begged him not to shoot the poor dinosaur, declaring herself to be a member of a group for the protection of wild birds. Nowadays, we know that this good lady was entirely correct to make such a request, since, as we saw earlier, birds are dinosaurs with wings and feathers.

We have to look at a news item in the *Times* of November 1919 to understand the origin of Captain Stevens's fantastic journey. According to the newspaper, a large animal had attacked a man called Lepage, who was in charge of laying a railway line in the Congo. After opening fire against the enormous animal, Lepage managed to escape. The animal subsequently attacked an indigenous village. The Belgian colonial authorities had decided to leave the animal alone because it was a "relic of the past." They supposed that the beast belonged to the group of the brontosaurs ("thunder lizards"), enormous quadruped dinosaurs with long necks and tails. On the basis of this news, a U.S. scientific institution was rumored to be offering $5 million for the capture of the dinosaur. It subsequently became clear that this news item was false, but Stevens did not know that when he set out with his dog to capture the animal. This story from 1919 is not an isolated one with respect to the legends of the persistence of dinosaurs in central Africa. Many European hunters and explorers collected such fables.

The survival of some type of nonavian dinosaur cannot be dismissed out of hand, improbable though it may be. Certainly there is still no unequivocal evidence of the existence of such an animal. The interpretation of the word "evidence" is one of the main lines of argument in what is known as "cryptozoology," a term first coined by the French researcher Bernard Heuvelmans, who bore the nickname of "the Sherlock Holmes of zoology." The main aim of cryptozoology, which sometimes went by the name of "romantic naturalism" in English-language literature, is the study of animals that have not yet been discovered. Naturally, this does not include all of them—it does not include, for example, insects and other invertebrates, which make up the majority of animal species still to be discovered. Cryptozoology concerns itself with terrestrial and aquatic vertebrates that are above all characterized by being situated on the boundary between legend, rumor, and reality. The aim is to try to disentangle in which of these areas the animal lies. Thus, creatures such as the great sea snake, the kraken, the yeti, and the Loch Ness monster are classic objects of cryptozoology.

The main line of argument of romantic naturalism runs as follows: we know that some living animals belong to groups that have scarcely changed over millions of years, and we also know that some groups of animals thought to have become extinct millions of years ago still have present-day representatives; therefore, it is probable that any *extinct* group, such as the dinosaurs, would appear in a barely changed form in the modern fauna. In the words of the late U.S. paleontologist George Gaylord Simpson,

> We know that some living animals belong to groups that have not changed greatly in millions of years. We know also that some groups believed to be extinct for millions of years have proved to have living representatives.

Therefore, it is likely that almost any "extinct" group, such as that of the dinosaurs, will turn up little changed in the modern fauna. That is generalizing from the plainly exceptional case, and that is where the scientist, although equally fascinated by the romance of his subject, parts company with the nonscientific romanticizer. It is a matter of judging probabilities. A negative cannot be proved in the full sense of the word. There is _some_ probability that there are little men on the far side of the moon, but the probability is infinitesimally small. For numerous reasons . . ., the probability that there are living dinosaurs is only a little larger. (_Natural History_, November 1959)

In his book, _Les Derniers Dragons d'Afrique (The Last Dragons of Africa)_ (1978), Bernard Heuvelmans recounts various legends and testimonies from that vast continent that could have been generated by the survival of dinosaurs. The most famous is the case of that known as Mokélé-mbêmbé, which supposedly inhabited the Congolese basin. It was a large animal, about the size of an elephant, with a long tail and neck, strictly herbivorous, with the habit of frequently retiring into underwater caverns. The animal, brownish gray in color, would have had smooth skin and would apparently have been similar to a small sauropod dinosaur in appearance. The U.S. researcher Roy P. Mackal carried out a study of the folklore of the local tribes to produce a description of the Mokélé-mbêmbé. His results differed markedly from the testimony gathered by Heuvelmans: the general appearance of the animal would have been similar, but it would have been between 5 and 12 m tall, with pink, green, or gray skin, a rooster-like crest on its head, and a horn on its nose. Its flesh would have been toxic.

It seems significant that this information (including the toxicity of its flesh) fits well with the sauropod dinosaur portrayed in the film _Baby—Secret of the Lost Legend_ (1985). _Baby_ was a production of the Walt Disney studios that adopted the worst defects of the mawkishness typical of Disney productions. A young U.S. couple, George (William Katt) and Susan (Sean Young), have met several months previously at the central African camp of a well-known paleontologist, Dr. Kiviat (Patrick McGoohan). Susan, a brilliant paleontologist who works with Kiviat, is summoned by the Red Cross because a local tribe has been poisoned as the result of eating the meat of a strange animal. Susan discovers that the animal is in fact a brontosaur, and she heads to the place where the discovery was made. Her husband joins her, and once they reach the exact place, they make friends with a primitive local tribe. Later, they find not just one dinosaur but a happy family: father, mother, and offspring, which they call Baby. Meanwhile, Kiviat, helped by a squad of soldiers, discovers the dinosaurs as well and manages to anesthetize the mother, although the soldiers kill the father. Susan and George escape from the place with Baby, who subsequently grows fond of them. Kiviat's team moves the female brontosaur in a large boat, and its cries attract the young dinosaur, who falls into the hands of the criminal paleontologist. Susan and George then set out to save the dinosaurs.

One of the most surprising aspects of *Baby—Secret of the Lost Legend* is the paleontologists' lack of amazement at discovering a surviving family of brontosaurs. The entire narrative has a sense of the realistic and the everyday that contrasts markedly with the majority of dinosaur films and that connects clearly with the cryptozoologic spirit. In addition, the film shows, in moral terms, the ambivalence that the appearance of a living dinosaur stirs in two paleontologists. On the one hand, Kiviat is an unscrupulous scientist whose only aim is glory at any price. Susan represents, at the other extreme, out-and-out honesty and ethics, a scientist who renounces everything for the happiness of a living being, even if it is something as exceptional as a dinosaur. Clearly, such a possibility in real life would not only mean fame for the scientist in question, but also the answer to so many questions that could not be answered from the fossil remains of the dinosaurs.

Without a doubt, the search for fame has been one of the most obvious driving forces behind the huge amount of research carried out in Loch Ness (Scotland), although, of course, we must also include those of human curiosity for the marvelous and the need we have for it. Everything seems to indicate that the legend of the existence of an enormous aquatic animal in Loch Ness goes all the way back to the 6th century. According to the tradition, in the year 565, the Irish missionary Saint Columba (521–597) had an encounter with the monster. Irrespective of that, the great celebrity achieved by the creature of Loch Ness had its beginnings in an article that appeared in the *Northern Chronicle* in August 1930. Three young people from the place observed the body of an animal some 6 m in length emerging from the water. The public commotion was highly significant. During the following years, there came testimonies, reports, and interviews in the media.

In the midst of this effervescence, the British film industry produced the 1934 film, *The Secret of the Loch,* a comedy in which a scientist is convinced that a monster inhabits Scotland's most famous loch. The animal comes from a prehistoric egg and is recreated, at the end of the film, using a lizard enlarged to giant size. A similar approach was taken in the U.S. film *The Crater Lake Monster* (1977). The story begins with a pair of scientists who have discovered some 1,000-year-old

Native American cave paintings. The paintings depict the clash between the Native Americans and an enormous plesiosaur (could this idea be based on the real fact that the mastodons in America became contemporaries of human beings?). While the paleontologists discuss the relevance of the discovery, a meteor falls into a nearby lake. The warmed water results in a plesiosaur hatching from a still-fertile egg. The monster sets about terrorizing and eating as many people as it can, but it is finally destroyed by an excavator.

After World War II, the craze about Nessie (the local name for the monster) revived notably. In 1951 and 1960, pictures of the animal were taken that many experts considered to be authentic. At the beginning of the 1970s, a private U.S. company, the Academy of Applied Science, assembled an extensive range of resources at Loch Ness, setting up underwater cameras with intermittent flashes. The result was some photographs supposedly of the fin of the animal and its neck and trunk. The former shows a vertebral appendix of an aquatic vertebrate similar to that of a plesiosaur, a reptile that became extinct 65 million years ago. The results of this experiment were published in the prestigious British scientific journal _Nature_. The animal was baptized as _Nessiteras rhombopteryx_, against the rules of the International Code of Zoological Nomenclature, since a new species cannot be formally proposed solely on the basis of photographs. Furthermore, the interpretation of the photographs was contested. According to some of the experts at the British Museum, the fin photographed might have belonged to a fish. There is currently a broad consensus concerning the other evidence that the neck, head, and body of the animal were in fact a rotten tree trunk.

At the end of the 1970s and beginning of the 1980s, apparatuses that used subaquatic echoes produced by particular sound sources (sonar, echolocation) began to be used in the search for Nessie. In 1982, a patrol boat fitted out with sonar equipment registered 40 contacts of notable strength and depth in comparison with those of known fish. In 1987, a major operation, known as Deepscan, was mounted. A total of 19 boats swept the loch bed in search of the monster, without any result. Whether Nessie really exists therefore continues to be questionable.

It has already been remarked that most people who believe in the existence of the Loch Ness monster claim that it is a plesiosaur. Cinema, however, has put forth other explanations. In Billy Wilder's masterpiece, _The Private Life of Sherlock Holmes_ (1971), the Loch Ness monster turns out to be the prototype of a submarine developed by the

British Empire's Secret Service. In an episode of the television series *Doctor Who*, "Terror of the Zygons" (1975), Nessie is an enormous cyborg that extraterrestrials, the Zygons, plan to use in order to take over the Earth. Returning to the popular idea of the plesiosauran nature of the beast from Loch Ness, if it really existed, then it would represent an exception to the great extinction at the end of the Cretaceous.

Intelligent Dinosaurs

What would have happened if the biotic crisis of 65 million years ago had not happened? One of the obvious predictions under these circumstances is that the dinosaurs would have continued to evolve to give rise to forms whose nature would be difficult to evaluate. Nevertheless, certain dinosaurologists have made the odd foray into this area. Such is the case of Dale Russell, who at the beginning of the 1980s produced the famous *Dinosauroid*. This creature is the end product of an intellectual exercise: the hypothetical appearance of an intelligent dinosaur from a theropod from the troodontid group. The result is a bipedal organism that stands completely upright and has a high degree of encephalization, similar to that of human beings. Russell's central idea has much in common with Ivan Antonovitch Efremov's aforementioned proposal in *Star Ships:* the humanoid form is convergent for any intelligent nonaquatic, terrestrial vertebrate. The dinosauroid therefore represents the best solution for the physical and physiologic problems imposed on an organism with a hypertrophied brain in a terrestrial environment. The form of human beings is well adapted and represents an objective "sought and solved" by natural selection.

A similar concept to Russell's dinosauroid was used by the U.S. science fiction author, Bob Buckley, in his story "The Runners" (1978). An expedition into the past, composed of a geologist, two paleontologists, and an astronomer-ethologist, observe the fauna of the end of the Cretaceous. The narrator discovers a pair of intelligent troglodyte dinosaurs that are capable of using rudimentary stone tools. The members of the expedition discover the existence of a supernova that, according to them, will be the cause of the extinction of the dinosaurs. In the end, moved by the great biotic crisis, they transport a number of

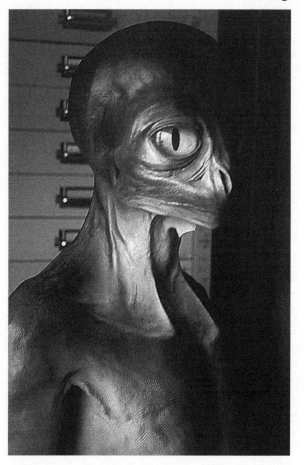

The paleontologist
Dale Russell's
dinosauroid.
Courtesy of the
Museo Nacional de
Ciencias Naturales
de Madrid (Spain).

eggs laid by the intelligent dinosaurs to their own time, in the belief that the species can be conserved in the future.

In recent years, several biologic thinkers, such as the famous U.S. paleontologist Stephen Jay Gould, have developed the idea that the course of evolution is unpredictable. The main reason is that the history of life is profoundly influenced by randomly produced phenomena, contingency factors that can bring about drastic changes in the historical succession of living beings. A good example of a contingency factor would be the Yucatán meteorite that occasioned the extinction of the nonavian dinosaurs. What would have occurred if the meteorite had not fallen? It is possible that the dinosaurs would have been able to give rise to intelligent, self-aware organisms of a form similar to the dinosauroid proposed by Russell. It is even possible that under such circumstances, human beings would never have appeared. In effect, under the domination of the varied dinosaur lineages, it is conceivable that our own zoologic class, the mammals, would not have diversified during the Tertiary, and primates would never have arisen. In T. Sullivan's humorous story, "A Dinosaur on a Bicycle" (1987), a civilized intelligent dinosaur (from the dominant race on the planet Earth) is shifted back in time in a pedal-propelled machine. In the Cretaceous, it encounters an endless array of chrononauts from different animal groups: simians (human beings), dogs, cats, and even cockroaches. The dinosaur time traveler thus realizes that there is no reason to continue to be proud of being a member of the dominant race on the planet because there was only the slightest probability (the whim of the universe) that dinosaurs, and not rats, grasshoppers, or dolphins, had attained self-awareness.

The television series *Dinosaurs* (1991) reveals a fantasy world before the appearance of "civilized" humans. The dinosaurs, which are intelligent, have established and organized a society for themselves. The main characters are the Sinclair family (which is probably a nod toward the U.S. oil company that has a sauropod as its logo), made up of the parents, two adolescent children, a baby, and a grandmother. The series is an entertaining reflection on human society and the human condition. The dinosaurs are used cleverly to highlight obvious social problems: exploitation in the workplace (father Sinclair has a huge *Triceratops* for a boss, who is continually pushing him around), discrimination against women (mother Sinclair has a colossal female sauropod as a friend who is prevented from working in her husband's company because a woman's place is in the home), and the problems of puberty (the Sinclair daughter's tail has not grown enough to attract the young boys). The series pays great attention to environmental problems: father Sinclair's company fells forests to build apartment blocks. The dinosaurs are irreversibly damaging their environment, which will lead to their extinction.

The fantasy story genre has produced intelligent dinosaurs on other planets and also in parallel dimensions. The series *Dinosaucers* (1987) describes the arrival on Earth of extraterrestrial dinosaurs that, equipped with advanced technology, try to save our planet from other, less friendly fellow animals. The degree of hostility toward Earthlings is at its greatest in the TV series *V* (1983–1985): intelligent lizards from another planet in the shape of human beings invade us with the aim of getting hold of water and food. The series made a great impact because of its political background and its true-to-life production values. The visitors are organized in a totalitarian society; they are Nazi-like aliens who efficiently use all the techniques of domination developed by the German National Socialists. James Blish (1921–1975) offered a different approach in his novel, *A Case of Conscience* (1958). A Jesuit biologist, Father Ramón Ruiz-Sánchez, is sent, along with other scientists, to the planet Lithia. The Lithians are huge, intelligent reptiles, more than 6 m long, and similar to dinosaurs. The Lithian society is perfect; there is no social inequality or crime, and the concept of God does not exist.

Our own space-time continuum is not the only possible one. Others are feasible, including those inhabited by intelligent dinosaurs. This is the idea behind films such as *Super Mario Bros.* (1993) and *Dinosaurs* (1990). In the former, dinosaurs continue to exist in a different universe to our own where their evolution has led to the appearance of self-awareness. The fall of a meteorite 65 million years ago that caused the release of a huge amount of energy was the factor that brought

A *Far Side* cartoon by Gary Larson. It spoofs films such as *King Kong* and *Godzilla*.

"Hey! Is that you, Zorak? . . . Small world!"

about a displacement of the space-time continuum. In *Dinosaurs*, three young boys are projected into a television cartoon series whose lead characters are intelligent dinosaurs.

The appearance in science fiction stories of dinosaurs endowed with the capacity of reason is yet another product of the anthropomorphization of animals, a reflection of the intense presence of the image of dinosaurs in contemporary societies. The anthropomorphization of dinosaurs is an indication of how they are considered to be friendly animals, which can at times stir feelings of tenderness and affection in those in the youngest strata of society. Nevertheless, these types of relationship between humans and dinosaurs are anomalous. The normal state of affairs is that dinosaurs and human beings face each other in a conflict, reflecting the traditional difference between humans and Nature.

19

The Coexistence of Humans and Dinosaurs

There are basically four types of relationship between humans and dinosaurs in the worlds of fantasy: (1) cordial, as in the case where the dinosaurs have been anthropomorphized (as in the case of the dinosaurs of *Dinotopia*); (2) affectionate, normally involving the treatment of dinosaurs as pets; (3) exploitative, with the human beings taking advantage of the dinosaurs; and (4) conflictual, confrontational relationships (the most common type).

The coexistence of children with dinosaurs is one of affection when they adopt them as pets. In the film *Dinosaurus!* (1960), a child be-

This is one of the original paintings from the book series *Dinotopia* by artist James Gurney. Original art copyrighted by James Gurney.

The artist James Gurney painted this picture of "Tree Town" from his series of books *Dinotopia*. This island is inhabited by both dinosaurs and humans, who have formed a joint civilization. In 2002, the first of the *Dinotopia* stories became a television miniseries produced by Hallmark. Original art copyrighted by James Gurney.

Tyrannosaurus rex, Triceratops, and human. This is a classic wall poster from the 1970s by legendary fantasy artist Frank Frazetta.

Card #10 from the cult card series *Dinosaur Attack!* (This series was a partial spoof of the 1950s cult card series *Mars Attacks!*) In this card the *"Rhedosaurus"* and the Giant Behemoth (both dinosaur movie stars) attack the Leaning Tower of Pisa in Italy.

friends a caveman, a contemporary of the dinosaurs, and climbs up on the back of a huge brontosaur with him. The Soviet author Kiril Bulychev was the author of the series of short stories collectively entitled *A Little Girl to Whom Nothing Ever Happens* (1974). The main character is Alice, a girl from the future. In the story "Brontia," the little girl makes friends with a live brontosaur. The dinosaur has developed from a frozen egg found on the banks of the River Yenisei.

The human beings have used dinosaurs as beasts for pulling carts or for riding, as beasts of burden, and even as machines of war. In the television series *Dinoriders* (1988), people from another planet arrive on Earth during the Mesozoic, fleeing from their ferocious enemies. Both bands use the dinosaurs as war machines, equipping them with artillery and armor plate. In Roger Zelazny's thought-provoking novel *Roadmarks* (1979), the Marquis de Sade plans to do away with a powerful industrialist of the 27th century using a suitable controlled *Tyrannosaurus*. Buster Keaton's film *The Three Ages* (1923) features a caveman who uses a sauropod in a much less tragic and violent way: as an animal broken in for riding. The 1977 film *The People That Time Forgot* is the sequel to *The Land That Time Forgot*. The story tells of the arrival on Caprona of a rescue expedition, flying over the ice in a hydroplane. After being attacked by a pterosaur, they make a forced landing. Then, in order to shift the aircraft, they attach it by a rope to the tail of a stegosaur, which obviously moves slowly and gently.

When it comes to it, dinosaurs can also serve as food for human beings. An ornithomimosaur is eaten by the survivors in the film *Planet of Dinosaurs* (1978). Arthur Conan Doyle describes the use of herds of *Iguanodon*, an ornithopod dinosaur, as livestock for primitive humans in his 1912 novel *The Lost World*.

Nevertheless, if we take a close look at the fantasy story, it appears that in most cases the interaction of one or more dinosaurs with human beings automatically gives rise to a relationship of conflict. In the following, chapter, I will examine this confrontation.

20

The Conflict between Humans and Dinosaurs

The conflict between humans and dinosaurs can be played out on three main stages, each of which has its own characteristics:

1. Natural synchrony of primitive humans and dinosaurs (the prehistoric tale).
2. Human beings move to the place where the dinosaurs survive.
3. Dinosaurs move to human societies (dinosaurs against civilization).

The prehistoric tale, as discussed previously, describes an archaic world populated by dinosaurs and primitive humans. The human beings find themselves subordinated by the giant beasts, and only their astuteness makes it possible for them to occasionally defeat the dinosaurs. The hero of the story of *One Million Years B.C.*, Tumak, manages to eliminate an *Allosaurus* by making the animal impale itself in the belly while running. In *Caveman* (1981), the ingenious Atook (played by former Beatle Ringo Starr) renders a tyrannosaur incapable of combat by feeding it a bunch of sleep-inducing fruit. It seems obvious that in the absence of ingenious human resourcefulness, the dinosaurs would be the dominant beings in this primitive world. The situations in which our distant ancestors watch a formidable struggle between two dinosaurs with a sense of terror and powerlessness are common in many of these stories.

When a group of explorers penetrates a lost world, the confrontation with the dinosaurs may be merely accidental, or it may be intentional. Stories in which dinosaurs become highly sought-after specimens of big game, some of which have already been mentioned, are paradigmatic examples of this second type. In the film *The Last Dinosaur* (1977), a refuge from Jurassic times is discovered beneath the po-

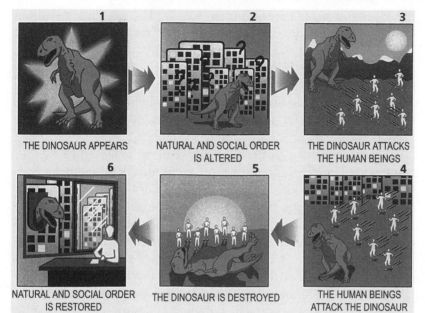

1	2	3
THE DINOSAUR APPEARS	NATURAL AND SOCIAL ORDER IS ALTERED	THE DINOSAUR ATTACKS THE HUMAN BEINGS

6	5	4
NATURAL AND SOCIAL ORDER IS RESTORED	THE DINOSAUR IS DESTROYED	THE HUMAN BEINGS ATTACK THE DINOSAUR

Basic structure of the confrontation between men and dinosaurs in the story of the "dinosaur against civilization" genre. Illustration by Gabriela Latapí.

lar ice. A U.S. multimillionaire with a passion for big-game hunting organizes an expedition to kill an enormous carnivorous dinosaur. In the end, some of the expedition members manage to make it back home, while the multimillionaire, an avid dinosaur hunter, finds himself up against the creature armed only with a rudimentary catapult. The story clearly underlines how the hunter and the last dinosaur are interchangeable, both signifying the same idea: they are the last representatives of obsolete, vanished worlds.

The appearance of a dinosaur in civilized human societies always means a serious disruption of order at various levels. These stories typically consist of four main stages: (1) the dinosaur appears; (2) the dinosaur attacks the human beings; (3) the human beings attack the dinosaur; and (4) the dinosaur is destroyed. It should be stressed that the increase in the degree of disorder that would result from the appearance of a dinosaur in the present day is not only a matter of social and natural stability, but also of the divine order. The breach of social order is obvious, and we can see its importance in films with titles that are variations on the "enormous dinosaur on the rampage in the city" theme. The urban destruction and the masses of horrified people are a clear symbol of this disruption of human society. The resulting assault on the natural order is also obvious: the dinosaurs had their opportunity in Nature and have no place in the present day. Finally, as we shall see later, a dinosaur may be perfectly identified with the myth of the monster as part of the instinct–reason conflict of the beast–human duality, a theme that borders on the supernatural. For all these reasons, it seems quite clear that the reestablishment of order, at every level, necessarily involves the obligatory destruction of the dinosaur.

How can a beast as powerful as a dinosaur be killed, and what is its significance? Evidently, in most cases, conventional weapons are powerless to put a stop to the threat posed by the huge beast. The solutions offered in the fantastic story belong predominantly to two areas: the use of a special weapon, or of the most accessible and conventional technology. Films such as _The Beast from 20,000 Fathoms_ (1953) and _The Giant Behemoth_ (1959) featured the first of these approaches. In the latter film, a dinosaur is awakened by nuclear tests. Although mortally injured by the radioactivity, it destroys the banks of the River Thames all the way to London. Finally, it is destroyed by radioactive torpedo launched from a minisubmarine. The _"Rhedosaurus"_ of _The Beast from 20,000 Fathoms_ is eliminated by a similar procedure: an elite sharpshooter manages to introduce a radioactive projectile into its body. Sometimes the solution is more imaginative, as is the case in the first of the films of the Japanese Godzilla saga, _Godzilla, King of the Monsters_ (1954). The formidable Japanese dinosaur-dragon is destroyed by a last-minute invention: an "oxygen destroyer" that eliminates free oxygen from Godzilla's aquatic environment.

In contrast to the special weapon approach, dinosaurs are on occasion destroyed by nonmilitary, and even conventional, technology. In the film _Dinosaurus!_ (1960), a U.S. builder is working on the extension of a port using explosives. The dynamite uncovers two frozen dinosaurs, which are dragged along the beach. The huge dinosaurs (a tyrannosaur and a brontosaur) are revived by lightning. The tyrannosaur devours several country folk and seriously wounds the brontosaur, which ends up sinking in quicksand. Meanwhile, the inhabitants of the island take refuge in the ruins of an old fort in order to confront the huge carnivore. They manage to repel it with fire, but that does not prove to be sufficient. In the end, the head of the U.S. company confronts the tyrannosaur with a mechanical excavator and thus manages to kill the animal, which falls off a cliff and sinks into the sea.

In the same way, the U.S. productions _Carnosaur_ (1993) and _Carnosaur II_ (1994) portray the destruction of the dinosaurs by the use of an excavator. Humanity is incapable of detaining the beast. But the hu-

Godzilla Gallery artwork from the card series based on the American-made movie *Godzilla 2000,* in which the beast now appears more lizard-like. (*left*) Godzilla is seen perched on top of the Chrysler Building in New York City. (*below*) The foot of Godzilla is seen stepping over a television cameraman in the streets of New York City.

Dinosaurus! (Universal, 1960). This film had a Neanderthal "caveman" resurrected along with a _"Brontosaurus"_ and a _Tyrannosaurus._

man beings match the claws and strength of the giant carnivorous dinosaurs with science and technology. The digger is nothing other than a
metal extension of the bones and muscle of the person controlling it. It
is an extension generated thanks to human science, which is what finally
does away with the beast. This concept can be extended to the development of sophisticated weaponry—even the stone ax in stories set in a
prehistoric background. Thus, the process of confrontation with the
dinosaur and its death have to be understood as a further triumph of
the human intellect over Nature, a theme that has run through various
mythologies since the beginnings of culture.

In the dinosaur–civilization conflict, there are two elements of outstanding importance: the army and the paleontologists. The military
are incapable of warding off the threat, clearly symbolizing the powerlessness of society in the face of the dinosaur's aggression. The paleontologists are the stratum put in charge of letting the army know what
measures are needed to get rid of the monster. They normally limit
themselves to describing some quirk of the dinosaur and stressing its
tremendous power and resistance to conventional weapons. The role
of the paleontologist is frequently almost that of a defender of Nature.
In effect, the paleontologists' voices are the only ones raised in defense
of the dinosaur's life—although, of course, out of purely scientific interest. The information about dinosaurs used by the paleontologists in
fantasy tales is similar to that of the general public. The second filmed
version of *The Lost World* (1960) includes a scene of an argument between Professor Challenger and Professor Summerlee about the survival of prehistoric animals. When the first of them announces that he
has seen dinosaurs, Summerlee asks him if they were big. Challenger
answers him angrily that he has yet to see a small dinosaur.

At some point, the paleontologist has been induced into the mythical gallery of the mad scientists of science fiction. This is the case of
Professor Bromley in the 1969 film *The Valley of Gwangi*. Bromley
possesses all three typical traits of the mad scientist (the paradigm of
whom is Dr. Frankenstein): he has no hesitation in cheating and lying
to achieve his ends (what he believes is science); he supports heterodox
hypotheses (the origin of humankind is much earlier than otherwise
believed, about 50 million years ago); and, finally, the English paleontologist will be destroyed by his own creature (in this case, Gwangi, a
carnivorous dinosaur). Paleontologists, like other people, are attacked
by the dinosaurs and are equally terrified of them. In fact, any human
being might experience a feeling of panic when confronting such huge
prehistoric beasts.

21

Dinosaurs and Terror

What are the fundamental things that produce a terror of dinosaurs in science fiction? The stories that tell of confrontations with present-day human beings are those that most clearly illustrate the monstrous nature of the dinosaur. The myth of the monster is above all based on the concept of abnormality, and the invasion of human society by a dinosaur is certainly a most abnormal phenomenon. The dinosaur symbolizes the unleashed and uncontrollable forces of Nature that humankind has not found out how to master and that humankind therefore fears. Many recent films overlook this symbolic subtlety and tell of earthquakes, volcanic eruptions, and floods equivalent in their destructive power to Godzilla destroying the port of Osaka. The offensive power of the giant beast is always terrifying. Ignorance of the exact nature of the threat posed by the beast gives rise to a degree of uncertainty, in the sense that nobody knows how to eliminate the dinosaur. We should remember that dinosaurs can be predators of human beings, against which humanity is not used to fighting.

The beastliness of dinosaurs is an attribute of their monstrous character, there being unequivocal evidence of this in numerous tales of the fantasy discourse. Films, such as *The Last Dinosaur* and the Japanese *Legend of the Dinosaurs and Prehistoric Birds* (1977), show dinosaurs as drooling, bloodthirsty animals whose claws and teeth are more than just a symbol of their cruelty to human beings. In the novel *The Land of Terror*, Doc Savage comes face to face with a carnivorous dinosaur, "the most terrible and repugnant monster ever seen by human eyes." Its fetid teeth are as long as a person's arm. Thus it should be stressed that the presence of a powerful meat-eating set of teeth in theropod dinosaurs bears a clearly terrifying significance in human culture.

The bestial image of dinosaurs may also have other symbolic con-
notations. Humans see in animals imperfect and disturbing images of
themselves. If one of the attributes of the beast is its violence, then the
symbolic capacity as a paradigm of violence is notably increased in a
giant beast. The beast is also a disproportionate representation of the
libido, for which reason dinosaurs have frequently been involved in the
classic myth of beauty and the beast. Above all, comics often illustrate
the hostile attitude of dinosaurs toward beauty (they apparently want to
eat it, although this attitude may represent other, more unconfessed
intentions). In the Judeo-Christian tradition, the beast is unfailingly

Dinosaurs and the "beauty and the beast" myth. The comic strip *Spirit of Africa*
by Marco and Beá, from the magazine *Rambla,* January 1984, García y Beá
Editores (Barcelona, Spain). Courtesy of Tomás Marco.

associated with evil, often in a supernatural context. So it comes as no surprise that the final destruction of certain dinosaurs, such as Gwangi and the "*Rhedosaurus*" from *The Beast from 20,000 Fathoms,* is brought about by the traditional purifying fire that eliminates the essence of evil.

The symbolic potential of dinosaurs is based largely on the image with which they have been recreated, especially in the cinematic medium. In the next chapter, the image of dinosaurs in the science fiction story will be examined.

22

What Dinosaurs Looked Like

Clearly, no human being has ever seen a living nonavian dinosaur. Their external appearance, posture, and attitude have to be deduced from the information provided by the fossil record. The process of reconstruction of the appearance of a living dinosaur passes through several stages. The first is to have a complete skeleton available in which the structural relationships between the elements (the location of the skull in relation to the series of neck vertebrae, the orientation of the bones of the limbs with respect to the vertebral column, arrangement of the tail vertebrae, etc.) are clear. Once it is reasonably certain that the posture in which the skeleton has been assembled is natural, the bones have to be covered with appropriate musculature, which begins with a consideration of the rough areas and projections surfaces of the skeletal elements. This then allows a dinosaur's muscular system to be reconstructed by comparing it with that of known vertebrates. Next, the external appearance of the tegument or skin (the possibility of skin folds and cutaneous projections) must be understood. In many cases, the texture of the skin of particular dinosaurs is known from impressions in the rock containing the skeleton, or even through fossilization of the teguments themselves. The final phase of the reconstruction of a dinosaur is to give it the appropriate skin color, although absolutely no information of this type exists. Therefore, the colors attributed to the dinosaurs, even in solidly scientific reconstructions, are merely speculative.

This degree of speculation is specifically recognized by the time traveler in the novel *The Dechronization of Sam Magruder*, by George Gaylord Simpson, published in 1996. Sam Magruder, a chronologist of the 22nd century transported to the Cretaceous, is unable to quickly identify the first dinosaur, a sauropod, that he sees. The reason is that

Dinosaur artist Stephen
Czerkas in his studio in
Utah. Many of his
sculptures are on display in
museums throughout the
world.

in reality, its colors are different from those with which it is usually
reconstructed. In fact, few modern-day artists would portray a sauropod
with an emerald-green body and brilliant red eyes, as Sam Magruder
discovers in his direct observation of the living dinosaur.

The appearance of dinosaurs in comic book stories and in science
fiction literature more or less fits in with the suggestions of paleontologists. The cinema does not follow suit, however; it is a medium in which
dinosaurs have taken on every type of appearance. This variety is probably related to the limitations of the material resources (special effects)
historically available in the film industry for reconstructing dinosaurs
on screen. Obviously, the most perfect dinosaurs, from the paleontologic point of view, have only been created since powerful computer
tools became available. The combination of paleontologic information available at any moment, the development of special cinematic
effects, and the budgetary constraints of any film have given rise to a
series of different types of dinosaur in the seventh art. In this way, an
initial division can be established separating the types of appearance of

Anatomical study for the reconstruction of the theropod dinosaur *Carnotaurus,* by Mauricio Antón. Courtesy of the artist and Grafismo, S.A.

dinosaurs into two categories: real dinosaurs and dinosauroids. In turn, the latter can be subdivided into three groups, which I call "paradino-sauroids," "sauriodinosauroids," and "dragodinosauroids."

Real dinosaurs are obviously those that are based on the paleonto-logic knowledge of the period that strive to achieve a certain degree of accuracy. Initially, there were essentially two cinematic procedures for producing this group of dinosaurs: stop-motion animation of life-size articulated models, and scale models. The U.S. prehistoric cinema of the beginning of the 20th century already used full-scale models. In

D. W. Griffith's film _Brute Force_ (1913), a rudimentary model of _Ceratosaurus_ threatens some primitive humans. The model could rock on its hind legs and open and close its mouth. In the 1926 comedy _The Savage_, a scientist finds a huge, friendly papier-mâché brontosaur on an island. Full-size models were used during the 1950s and 1960s, but since the 1980s, expectations have been heightened as a result of developments in mechanical engineering and robotics. Currently, one of the greatest exponents of this field of work is the U.S. specialist Stan Winston, who was jointly responsible for the dinosaurs in _Jurassic Park_ (1993). The mechanical dinosaur models produced by Winston's team are equipped with linear potentiometers that exactly reproduce some 40 different movements controlled by an operator who manipulates a small-scale model.

Without a doubt, the process of bringing dinosaurs to the screen that has the most numerous and stalwart supporters is that of the scale models animated using relatively simple camera methods. This technique begins with the creation of an internal articulated skeleton, normally of metal. This base is covered with latex and other flexible materials to give the final figure an appropriate external texture that imitates

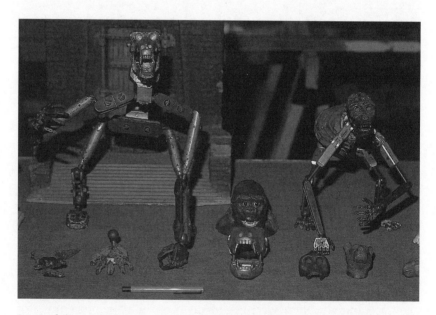

Some of the original armatures built by Willis O'Brien and Marcel Delgado for the films _King Kong_ (RKO Pictures, 1933) and _Son of Kong_ (RKO Pictures, 1933). These models are now in the private collection of Bob Burns. Photograph by M. Brett-Surman.

the skin of a dinosaur. Stop-motion animation uses the same principle as in cartooning, but in three dimensions, in the manipulation of these small-scale models. The operator slightly moves the figure (its legs, body, neck, etc.) and shoots a single frame. This action is repeated thousands of times, and the speed of projection of the film (24 frames per second) produces the illusion of movement.

As mentioned before, this animation technique was developed at the beginning of the 20th century by Willis O'Brien, although its indisputable master was Ray Harryhausen, creator of the dinosaurs in many films, including *The Beast from 20,000 Fathoms* (1953), *One Million Years B.C.* (1966), and *The Valley of Gwangi* (1969). Harryhausen's universally recognized work was based on the attempt to combine great realism in the construction of the dinosaurs with convincing movements and postures. Other masters of frame-by-frame animation are Jim Danforth, who made *When Dinosaurs Ruled the Earth* (1969) and *Caveman* (1981), and Phil Tippett, who was in charge of creating the dinosaurs for *Jurassic Park*. All these specialists are characterized by the similar course of their careers, which began with a personal fascination with the world of dinosaurs. Some of them, such as Stephen Czerkas, the creator of the models seen in *Planet of Dinosaurs* (1978), even publish research articles in professional paleontologic journals.

Dinosaur paleontologist M. K. Brett-Surman playing with an *Allosaurus* model built by artist Stephen Czerkas.

The majority of the small-scale models built for use in the process of stop-motion animation have tried to be faithful to the paleontologic knowledge as it was at the time the films were made. The theropod dinosaurs, from O'Brien's times to the works of Harryhausen at the end of the 1960s, were characterized by an external appearance featuring enormous scales. One of the most famous carnivorous dinosaurs in the history of cinema is the _Allosaurus,_ which dies, impaled, in _One Million Years B.C._ (1966). Vigorous and active, this theropod, measuring almost 3 m in height on screen, is constructed with enormous dorsal and cephalic dermal plates and a series of longitudinally oriented ventral scales in a form similar to that of snakes. The movements of this theropod are generally convincing.

In an interview in 1974, Harryhausen pointed out the impossibility of knowing how dinosaurs actually moved. In the case of the carnivores, Harryhausen doubtlessly opted for an avian model, which gave excellent results. Nevertheless, there is an additional problem: the tail. In general terms, the movement of a large bird, such as an ostrich, can be taken as a reference model for that of a giant theropod. However, no bird has the large caudal appendage of an allosaur or a tyrannosaur. Today, it is believed that this appendage carried out an important function in helping a bipedal dinosaur to walk, by moving from left to right to provide the correct balance while it stepped forward. In Harryhausen's and other creators' theropod dinosaurs, such as Marcel Delgado's allosaur in _The Beast of the Hollow Mountain_ (1956), the tail represents an additional movement that seems impossible, given the rigid skeletal structure of the caudal series in the theropods. The movement of the tails of these cinematographic dinosaurs amounts to a nervous wave halfway between the lateral movement of a snake and the swishing of a cat's tail.

To date, the dinosaurs of the cinema with the most accurate movements and postures are those produced by the computer-generated images in _Jurassic Park_ (1993), _The Lost World: Jurassic Park_ (1997), and _Jurassic Park III_ (2001). The first step consists in making a scale model that is digitalized in three dimensions. The result is a network of lines (known as a wireframe) that reproduces the gross appearance of the animal. Upon this digital image are superimposed the skin texture and a complex system of tegumental wrinkles and folds, which must also move in an appropriate way as the animal walks. One of the many advantages of this system is that each dinosaur can be duplicated an unlimited number of times, thereby allowing the generation of such memorable sequences as the stampede of the herd of _Gallimimus_ in

Jurassic Park. The theropods in this film move in an obviously avian manner.

Harryhausen's intuition of the 1960s, the intimate relationship between dinosaurs and birds, is now firmly supported by the conclusions of present-day dinosaurologists. These ideas, increasingly firmly established as part of general knowledge, are not confined to the cinema alone. In the 1992 novel *The Ugly Little Boy,* by Isaac Asimov and Robert Silverberg, a private company from the 21st century carries out time probes. By this process, animate and inanimate objects can be collected from the remote past and brought forward in time. In this way, they transport a group of trilobites from 500 million years ago and a Neanderthal child from 40,000 years ago, who is the main focus of the story. Among the recovered items is a small dinosaur that is described in the story as being a species of long-tailed, scaly chicken. There is no doubt that it is really a small theropod remarkable for its tiny, dangling arms and its hands, which open and shut convulsively. The head is similar to that of a bird, with brilliant scarlet eyes and an electric blue cephalic crest. It is bright green in color, with darker stripes.

Returning to the dinosaurs of *Jurassic Park,* although they are highly accurate, there are one or two peculiarities that should be mentioned. For example, let us consider the case of *Dilophosaurus.* This primitive theropod was some 3 m long in real life. However, it was reduced to half that length in the story, where it also has an extendable cervical collar and the ability to spit poison, which are traits for which there is no evidence (although this would be practically undetectable, particularly in the case of the poison) in the fossil record. The case of *Velociraptor* is also peculiar, but in a different way. Michael Crichton conceived these theropods in their real size, equivalent to human beings. This characteristic allowed these dinosaurs to be more threatening, since they could chase humans inside buildings. Against the opinion of some of his advisors, Spielberg insisted on increasing the size of the velociraptors on the screen, fearing that they would be unable to fulfill their role as mortal enemies of humanity. Interestingly, during the summer of 1992, a velociraptor was found in the east of Utah (United States) of a similar size to those in Spielberg's film. This theropod, from the Lower Cretaceous (125 million years ago), has been named *Utahraptor.*

A clearly inadequate method for attempting to represent real dinosaurs on the screen consists of putting an actor into a rubber suit. One of the first films that took this approach was *One Million B.c.* (1940), thus giving life to the same character that Harryhausen would create, using a frame-by-frame animated model, in the second version of the story, _One Million Years B.C._ (1966). One of the most recent films to use the risky rubber suit procedure was _Baby—Secret of the Lost Legend_ (1985). In this example, the procedure is complicated by the fact that _Baby_ tells the story of a family of sauropods. As we know, sauropods are quadrupeds, and so the person with the responsibility for giving life to the animal had to bend over and go on all fours. The resulting recreation of the dinosaur is rather unconvincing. An interesting additional anecdote is that the sauropods recreated in _Baby_ had the position of their nasal openings altered, from being dorsally situated to a point at the end of the snout. According to the people responsible for the film, this mammalian readjustment made the little dinosaur more familiar to human eyes, and so therefore they had a greater capacity for transmitting favorable emotions to the spectators. In 2001, Larry H. Witmer has proposed, after painstaking comparative anatomy studies, that the position of the nostrils of the sauropods would have been intermediate, between the dorsal position of the classical model and the position proposed for Baby.

Up to this point, I have looked at various aspects of real dinosaurs on screen. As the reader will recall, the next category that I am going to consider is that of the dinosauroids. This term is proposed for those dinosaurs in cinema whose appearance clearly differs from that suggested by the available paleontologic knowledge. Within the dinosauroids, the first group is that of the paradinosauroids. These are big beasts whose appearance is like that of particular types of dinosaur and that, in reality, are built from a mixture of characteristics of real dinosaurs, especially theropods and sauropods. Typical paradinosauroids include the enormous radioactive animal in _The Giant Behemoth_ (1959), the affectionate mother of _When Dinosaurs Ruled the Earth_ (1969), and the marine creature of _The Beast from 20,000 Fathoms_ (1953). According to Harryhausen, the appearance of the latter, which is called a "_Rhedosaurus_" in the film, was in response to the need for the tale to have a new dinosaur that nobody could recognize, a creature of uncertain origin, although one well known to the paleontologists in the film, the result of the explosion of an atomic bomb. Paradinosauroids are quadruped animals similar to sauropods but with much shorter necks and with powerful heads akin to those of the great carnivorous dinosaurs. This combination of characteristics might perhaps provide some idea of the origin of the paradinosauroids. In the first place, we would have to admit that they have been designed to be dinosaurs guaranteed to be convincing in the eyes of the spectator. Second, they must be giant beasts of proven ferocity. These two points lead to the necessity of combining the great biomass of a sauropod with the aggressiveness of a carnivore. In any case, it seems likely that Harryhausen's *"Rhedosaurus"* was subsequently a source of inspiration for other paradinosauroids.

Among the cinematic dinosauroids, there is a truly special group—the sauriodinosauroids. These consist of innocent lizards (iguanas and monitor lizards, among others) and baby crocodiles to which rubber horns, spines, and dorsal fins are attached. This procedure, usually utilized in low-budget films, was used repeatedly, especially in productions in the 1950s and 1960s. Among the more significant titles, worth mention are *One Million B.C.* (1940), *King Dinosaur* (1944), *Journey to the Center of the Earth* (1959), and *The Lost World* (1960). This procedure has few adherents, and in several instances, it is based on incorrect basic suppositions. The idea is that present-day lizards are closely related to dinosaurs—or, to put it more bluntly, that the lizards of our countryside are no less than shrunken, pale shadows of former dinosaurs, relict forms of an ancient, lost splendor, and therefore have every right to portray their enormous disappeared relatives on screen.

As well as the great phylogenetic distance between lizards and dinosaurs, which implies obvious differences of general structure, another argument can be advanced to cast doubt on the adaptation of lizards in order to pass them off as dinosaurs. One of the most obvious characteristics of dinosaurs is the arrangement of their legs. The femur is made vertical in a similar fashion as occurs in mammals. This arrangement is an undoubted improvement in the dinosaurs' locomotory capacity. In lizards and crocodiles, however, the femur is oriented obliquely relative to the body. This arrangement results in a peculiar type of movement, different from that which dinosaurs would have had. For these reasons, it is tough on the dinosaur fan when, for example, in the film of *The Lost World,* Professor Challenger, an eminent paleontologist, identifies a monitor lizard as a *Brontosaurus* and another disguised lizard as a *Tyrannosaurus.*

The term "dragodinosauroids" has been proposed for the final group of dinosauroids. All of them are characterized by being created by an actor hidden within an ostentatious rubber suit. These are the formidable creatures of Japanese cinema, which have made possible the development of one of the richest and most complex strands of mythology within dinosaur cinema.

23

Japanese Creatures

The Japanese films featuring giant monsters, a highly popular genre in Japan, are known by the name of *Kaiju Eiga*. The genre is basically targeted at the youngest of spectators, although it has a wide acceptance among the most varied strata of Japanese society. The monsters of *Kaiju Eiga* are typically combinations of various types of animals. Many of them are of prehistoric origin, although, as we shall see later, they may arise by a multitude of other factors, both of terrestrial and extraterrestrial origin.

The first Japanese film featuring giant monsters was *Godzilla, King of the Monsters* (1954). The version shown in the United States and Europe contained additional scenes, which included a U.S. journalist, witness to the story in which eight ships are sunk in Japanese waters. The few survivors have strange burns. One night, a village on the island of Oto is destroyed. Its inhabitants insist that the catastrophe was caused by an enormous legendary creature, Godzilla. ("Godzilla" is the Americanized version of the Japanese word "Gojira.") Later, the scientists discover increased levels of radioactivity in the destroyed village and also a living trilobite (an extinct arthropod from the Paleozoic era). An enormous head suddenly appears from behind a hill. The monster returns to the sea, leaving the impression of its feet and tail on the beach. The animal is identified as being from the Jurassic, an intermediate between terrestrial and marine forms, 130 m tall, the result of the atomic bomb tests. Godzilla appears in the Tokyo bay and destroys the Japanese capital city. It will finally be annihilated by a new weapon, the "oxygen destroyer," which reduces it to a skeleton.

One of the essential premises of the *Kaiju Eiga* is the constant return of the monsters, even though they were apparently destroyed in

Godzilla, King of the Monsters (Toho, 1954).

the previous film. In this way, although Godzilla was eliminated in its first cinematic outing, it turns up again, without any explanation, in a second film, *Gigantis, the Fire Monster* (1955, and retitled for its release on video in the United States as *Godzilla Raids Again*). The story begins with a strong warning against the nuclear arms race. Two hydroplane pilots who work in seeking out shoals of fish discover two monsters fighting on an isolated island. One of them is Godzilla, the other a species of spiny quadruped, a cross between a wolf, a dragon, a hedgehog, and an ankylosaur, called Anzyllas or Angurus, and a number of other similar names. The two huge animals fall into the sea. In the end, Godzilla appears at night in the Osaka bay. The authorities black out the lights, and a squadron of airplanes fires flares out to the open sea. This activity attracts the monster, and it moves away from the coast. However, an accident brought about by the flight of some criminals leads to a major fire in the Osaka harbor, attracting Godzilla once more. Anzyllas also shows up in the harbor and begins to fight furiously with Godzilla, who then kills the spiny adversary and swims away. The city of Osaka has ended up being almost completely destroyed. Godzilla later sinks a Japanese fishing boat, and the entire planet sets off in search of it while it is on a frozen island. There, within a high-walled corry with a single access route, a squadron of Japanese fighter planes attacks it. Faced with the impossibility of destroying the monster with conventional weapons, the pilots bombard the icy walls of the corry. In the end, Godzilla is buried beneath avalanches of tons of ice.

Since this second film, Godzilla has appeared on many occasions— in at least 13 titles up to the 1970s, and in various recent productions: *Godzilla 1985* (1984), *Godzilla versus Biollante* (1990), *Godzilla versus King Ghidorah* (1991), *Godzilla versus Mothra* (1992), *Godzilla versus Mecha-Godzilla* (1993), *Godzilla versus Space-Godzilla* (1994), *Godzilla versus Destroyer* (1995), and *Godzilla 2000* (2000). Another characteristic peculiar to the *Kaiju Eiga* is the interaction between the different monsters. Thus, Godzilla has confronted or has allied itself with a vast range of other gigantic creatures. A brief list would include Rodan, an enormous pterosaur that arose from the depths of a volcano; the aforementioned Anzyllas; King Kong (courtesy of the U.S. film industry); Mothra, an enormous moth that can appear in its adult form or as a huge caterpillar; Ghidorah, a three-headed extraterrestrial beast with membranous wings, whose weight has been calculated by some experts as being 30,000 tons; Ebirah, a species of gigantic lobster; Spiga, an enormous spider; Manda, a giant water snake with rudimentary limbs; Gigan, a bipedal extraterrestrial monster with massive claws and rudimentary wings; and Megalon, a species of beetle more than 100 m tall from the mythical continent of Mu. It should be pointed out that all these creatures were produced by the Toho film company, and that other companies have other monsters (a procedure analogous to the golden years of the Hollywood star system), as in the case of Gamera, an enormous tortoise that can fly like a flying saucer; Guilala, an extraterrestrial biped with antennae and lateral processes on its skull; and Gappa, a species of fabulous winged dinosaur that can fly and swim at great speeds.

Godzilla and the other monsters are the result of atomic tests. We should not forget that Japan is the only country to have suffered nuclear destruction by bombing, in August 1945. This dramatic event had hugely significant consequences in all spheres of Japanese life and culture. It is thus clear that, in part, the giant monsters embody the atomic threat to which the Japanese people are particularly sensitive. However, as has been remarked in chapter 21, Godzilla and company also represent the threat of natural disasters. The Japanese tradition maintains that the natural order must form the basis of human institutions: the social order must be preserved under all circumstances. The giant monsters of Japanese cinema are born transgressors of these ideas of the political order. Their impact on Japanese society is clearly greater than on any other culture that has a lesser sociopolitical sense. The fact that Godzilla and Gamera are equivalent to a tidal wave or volcanic eruption immediately explains two important characteristics of the *Kaiju Eiga*: first, the monsters are invulnerable and human beings can never destroy them (at least with conventional weaponry), in the same way that nobody can fight an earthquake; and second, the monsters reappear recurrently and unpredictably, razing everything in their path like a river of lava or a tornado.

In contrast to this characteristic of threat (as a radioactive or a naturally destructive phenomenon), the giant Japanese creatures sometimes play a different role: that of defenders of humanity. The most obvious case is of Godzilla, who changes from being an evil monster in the first films to allying itself with humanity against particular sources of aggression, such as extraterrestrial invasions and even the pollution generated by human beings themselves. Godzilla struggles against the monster Hedorah, an enormous, shapeless mass stemming from worldwide pollution in *Godzilla versus Hedorah* (1971). Thus, Godzilla represents a superhero defender that Japan offers to the world. (Some commentators have interpreted this attitude as being a kind of Japanese "compensation" for having been one of the countries responsible for unleashing World War II.) It also seems clear that after such a long-running saga, Godzilla today represents a national Japanese symbol and a projection of Japanese vitality at every level.

Godzilla's origins are explained in *Godzilla versus King Ghidorah* (1991). In 1944, the garrison of the Japanese army on a Pacific island is saved from extermination by the appearance of a dinosaur that puts the U.S. army to flight. The grateful Japanese troops bestow military honors on the dinosaur, which later turns into Godzilla as a result of the nuclear tests carried out in the Bikini Atoll. The film has a flavor of nationalist fervor in which Japanese economic might and future planet-wide preponderance are emphasized. In a way, Godzilla represents the Japanese national spirit in the film, the symbol of power of an empire that has only just begun to wake up, despite the recent Japanese economic crisis.

In short, Godzilla moves in the ambiguous moral sphere of the eternal duality of good and evil. There have been several attempts of varying credibility to explain this ambiguity. We should not forget that Godzilla represents on the one hand a clear threat, whereas on the other, it embodies the best of the oriental tradition of the dragon myth, which, as we know, basically represents supernatural powers that are favorable to humankind. Godzilla and other monsters also present a clear ambiguity with respect to the instinct–reason (beast–human) dichotomy. Throughout the films, the degree of Godzilla's anthropomorphization increases in proportion to the extent of its alliance with humanity, even to the point where it communicates telepathically with other monsters.

Finally, another factor of great significance in understanding the phenomenon of the *Kaiju Eiga* is Shintoism, the traditional religion of Japan, which is a set of beliefs and practices that existed in Japan before the arrival of ideas from China and India. Shintoism is a pantheistic religion whose countless gods are family ancestors and, characteristically, any phenomenon or object of Nature: the sun, the rain, the wind, the mountains, and all manner of animals—such as wolves, foxes, snakes, birds, deer, monkeys, insects. All these deities are responsible for the appearance and faculties of the giant monsters of Japanese cinema. One of the best examples illustrating the relationship between Shintoism and the *Kaiju Eiga* is the popular television series *Ultraman* (1966). In the episode entitled "Freeway 87," the spirit of a child who died in a traffic accident conjures a prehistoric winged reptile (which emerges imposingly from within a mountain). The massive beast attacks the vehicles traveling on the freeway where the child died. In another episode, some school pupils, searching for fossils in an outcrop discover a huge dinosaur (a "*Stegodon*"). The animal, which is found in a living posture with its four legs in a vertical position, as would delight an expert in taphonomy, still has its vital fluids intact, despite its skeletal appearance. The children stop the workers, who are constructing a public building, from destroying the dinosaur.

The *Kaiju Eiga* are closely associated with the oriental dragon. This and the eastern dragon tradition are intimately bound up with the roots of the myth of the dinosaurs in modern societies.

24

Dragons and Dinosaurs

If we look at the wide range of dragon iconography in the Judeo-Christian tradition, we can see surprising similarities between the image of dragons and dinosaurs. Is there any direct relationship between the two? In other words, could the discovery of the remains of dinosaurs in our earlier history have given rise to the dragon myth? In Europe, most of the dragon legends that can be related to paleontology are based on fossil remains of Quaternary mammals. Something similar has happened in China—in this case, bones of mammals from the Cenozoic age were found. Some authors have expressed doubts about the origin of particular dragons in the European iconography, which may be based on the discoveries of remains of plesiosaurs, as will be discussed later.

A number of European caves, especially in central Europe, traditionally maintain the name of the dragon's (or dragons') cave or lair. During the 17th century, the German doctor Petersonius Hayn found some large skulls, isolated teeth, and bones in several caves in the Carpathian Mountains around Moravia. In 1673, Hayn had an article published by the Halle Academy of Sciences entitled "Skulls of Dragons in the Carpathians." Around the same time, another German, named Vette, found similar remains in Transylvania. According to their discoverer, these bones belonged to flying dragons. Illustrations of the material described by Hayn and Vette still exist. In both cases, they are of Quaternary cave bears, a powerful animal that was one of the largest carnivorous mammals. The famous Austrian paleontologist Othenio Abel analyzed the legend of the dragon of Klagenfurt (Austria) early in the 20th century. At the beginning of the 14th century in the spot known as the "Dragon's Grave," the skull of a woolly rhinoceros from the ice age was found and was subsequently exhibited in the city's town hall. This specimen served as a model for the sculptor Ulrich Vogesland for his creation of a statue of a dragon, which today is an emblem of the city of Klagenfurt.

Some German paleontologists from the 19th century admitted that it was the famous complete skeletons of plesiosaurs from the Jurassic sites in the south of Germany that had given rise to the local dragon legends. In fact, the traditional dragon iconography of this central European area suggests interesting similarities between dragons and plesiosaurs with respect to their size and morphology. One of the best examples of this was described by the Jesuit Athanasius Kircher in his book *Mundus Subterraneus* (*Subterranean World*, 1678). The work describes, with a large dose of inventiveness, the marvels that lived in the gloomy world of caves, caverns, and underground galleries. Kircher tells the story of the celebrated dragon slayer Winkelreidt, who succeeded in killing a dangerous specimen from near the village of Wyler (Switzerland). The proportions of the monster are broadly reminiscent of those of a plesiosaur. The fins of the plesiosaur could have been interpreted as being wings.

From time immemorial, Chinese pharmacopoeia has used so-called dragon teeth (*long chi*) and dragon bones (*long gu*). This remedy, still used to this day, relieves maladies of the kidney, digestive system, and liver, and is also used to treat constipation, epilepsy, nightmares, and dysentery, among other complaints. The bones and teeth of dragons are a commercial product in great demand. The Chinese collect these materials from caves and other sites with stratified sediments. The usual interpretation is that they come from dragons that could not rise into heaven because of the lack of clouds or rain. In fact, most dragon bones are from Miocene mammals from northwest China. Nevertheless, an ancient Chinese example is known that might represent a direct relationship between dragons and dinosaurs. A text written at the end of the 3rd and beginning of the 4th centuries describes the discovery of dragon bones in Wucheng in the province of Szechwan. It explains the phenomenon in a similar way to the popular interpretation mentioned above. A dragon that was trying to ascend from a mountain into the heaven found that the gate to it was closed. It fell back to Earth and was buried in the subsoil. This evidence seems particularly significant given the well-known richness in dinosaur bones of the Chinese province of Szechwan. If one day it were possible to prove that the bones of the ancient Chinese text were really dinosaur bones, this would signify two things: first, the oldest known reference to these animals; and second, the only evidence found so far of a direct relationship between dino-

Dinosaur bones are obviously found in the subsoil, within sedimentary rocks. They can be immediately identified by their appearance as being of organic origin. Nevertheless, other attributes clearly relate them to the inorganic world of rocks and minerals. Because of this duality, which is common to the majority of fossils, magic or mystical properties are often conferred upon them. The picture illustrates a fragment of the femur of _Iguanodon_ from the Lower Cretaceous of Buenache de la Sierra (Cuenca, Spain). The medullar cavity of the bone has been filled through one of the surfaces with calcite (calcium carbonate) crystals, although remains of the spongy bone tissue that constituted the interior of the enormous femur can still be made out.

saur remains and the dragon myth (remembering that the plesiosaurs, the possible inspiration for some European dragons, are not dinosaurs).

Even so, it would not be very rigorous to limit the explanations of the rich dragon mythology only to the possible relation between the fossil remains of animals from the past and the legends of various dragons. The entire iconography of the dragons from any era and culture is connected with complex symbologies stemming from magic, religious, and even alchemical beliefs, as in the case of the Western dragon. In many cases, what primarily determines the nature of the dragon is entirely a cosmologic–religious conception, as is the case, for example, of the famous Aztec god Quetzalcóatl, who is typically represented as a feathered snake. The snake symbolizes the Earth and physical matter, whereas the feathers, the bird, symbolize the spirit. Quetzalcóatl is one of the principal divinities of the various Mesoamerican cultures, the lawmaker, inventor of the calendar, and pioneer of corn cultivation. It is also believed that he was the founder of the practice of bloody sacrifices, at first involving the blood of a penitent and later the heart of the victims, which was torn from the chest with an obsidian knife. This bloody rite was shown in the film *Q, The Winged Serpent* (1982), along with the destruction of the monster, which had built its nest in the vault of the Chrysler Building in New York. In 1945, the U.S. film *The Flying Serpent* told the story of a mad archeologist who murdered his victims with the help of the Aztec god Quetzalcóatl.

On the other hand, the dragon myth might not be connected with fossil remains of animals from the remote past, or with complicated magical concepts, but rather with large beasts that are contemporaries of humanity. Indeed, it is known that in a number of contexts, "snake" and "dragon" mean the same thing. Additionally, the relationship between dragons and crocodiles, which comes from antiquity, seems beyond doubt. Some Greek texts describe the dragon with the customs, appearance, and dimensions of crocodiles. Therefore, it is not surprising that on many occasions, the Spanish conquistadors took alligators and American caimans to be dragons. The equation of crocodile = dragon = dinosaur persists in a number of modern-day stories, such as the U.S. film *Alligator* (1980). A young alligator grows to a size of more than 10 m in the sewers of a large city, until it decides to come up above ground.

Independent of the possible material origin—biologic or cultural (magic, religion, etc.)—of the dragon legends, the myth of the Western dragon has specific characteristics, which will be examined shortly. The basic aim is to compare the surprising similarities between the structure of the dragon legends and the stories of the dinosaurs produced by the human intellect within the framework of science fiction.

Some of the elements constituting the collective unconscious of present-day humanity are similar to particular ideas created by ancient humans. This phenomenon was termed "archetypal images" by Carl Gustav Jung. Many of the animal symbols in our society are archetypal images, created by ancient magical or religious traditions. Modern advertising that associates a powerful automobile with a lion is associating this cat with an archetype: several cultures in the Middle East, in Judea and Mesopotamia, identified the lion with strength and power. In the same way, we can say that dinosaurs currently represent the archetypical image of dragons.

A number of European tales place the dragon in an underground world. The conflict with human beings arises when a dragon, lost in a maze of caves and galleries that make up its habitat, appears on the surface. The relationship with the dinosaurs is obvious: their remains are always found buried, and they belong to the unknown, underground world. An observer poorly informed about biology and taphonomy would come to the immediate conclusion that the dinosaurs were underground inhabitants.

Dragons, normally associated with a cave that is the connection with their world, typically defend material treasures or their possession of a maiden. In the stories of the lost world, the dinosaurs commonly act as aggressors against those who wish to carry off the fabulous treasures (e.g., gold, diamonds). Further, many dinosaur stories tell of encounters with maidens within the context of the extended beauty and the beast myth (see chapter 21). As well as defending their material or virginal treasures, the dinosaurs guard within themselves a more subtle treasure: scientific knowledge. The attraction felt by humans for the lost world is born primarily of this intellectual interest. It is the desire to understand dinosaurs better that drives _The Lost World_'s Professor Challenger and others like him to explore the lost world.

Dragons are typically associated with the magical and supernatural world. Dinosaurs also feature in a number of such stories. One of the most characteristic examples is that of the carnivore that is burned to death in a cathedral in _The Valley of Gwangi_. In the film _Dr. Mordrid_ (1992), two extraterrestrial characters with supernatural powers (good and evil, of course) end up in a showdown in the great hall of a natural history museum. The evil being gives life to an enormous _Tyrannosaurus_ skeleton, which attacks the police. To avert the evil, the beneficent extraterrestrial brings a large mastodon skeleton to life. The story "Green Brother" (1982), by Howard Waldrop, describes a Native American boy's search for a vision that will allow him to reach the age of majority. This is a matter of a mystical connection with the spirit of an animal. The animal in question is a tyrannosaur, and the boy knows that its bones are located in land near a fort occupied by the troops against which his tribe are at war. A magic spell allows an enormous dinosaur to come back to life, which then attacks the white men.

Finally, one of the characters most typically associated with the dragon mythology is the figure of the dragon slayer. This character has been widely used in the Christian tradition and in many other religious, mystical, and magical spheres. The paradigmatic example within Christianity is Saint George, who, according to the devout tradition, destroyed a dragon, a symbol of the dark forces of hell—once again, the subterranean world. One of the best-known dragon slayers not from Christianity is the hero Siegfreid, from *The Nibelungen,* an epic German poem written around 1200. Siegfreid kills the dragon and accidentally drinks its blood. As a result, he is able to understand the language of the birds (avian dinosaurs). This aspect of the story of the struggle of Siegfreid against the dragon illustrates another common characteristic of dragons, their association with Nature, by which they are both protectors and symbols of it. In this respect, it should be emphasized that the dinosaurs, as is obvious, are essentially neither protectors nor symbols, but rather part of Nature itself. In the context of their integration within Nature, dragons can be considered from more biologic perspectives.

This is what happens in the film *Dragonslayer* (1981), a magnificent tale that describes a pre-Christian medieval society in which the controlling power of magic plays a clear role. However, in fact, the dragon will finally be destroyed by a magician who shows that he has an empirical knowledge of dragons (he identifies an isolated tooth and works out the age of the dragon from one of its scales). We should remember that the information provided by paleontologists about the living habits and other characteristics of dinosaurs are of vital importance for their destruction in science fiction stories. Thus, the function of the magician and the paleontologist are perfectly comparable.

In the next chapter, the information that the fantasy story offers us about dinosaurs' way of life will be examined, but first let us reflect for one last time on the intimate connection between dinosaurs and dragons. A common sociocultural question concerns the reasons for the popularity of dinosaurs (see chapter 8). The fact that the dragon myth has been substituted by dinosaurs might explain, at least in part, human beings' fascination for these animals. Thus, dinosaurs would have been firmly rooted in the collective consciousness of humanity since remote epochs, including those before antiquity.

25

The Dinosaurs' Way of Life in the Fantasy Story

Many of the unknowns in current dinosaur research are concerned with their living habits. Where did they live? Did they defend a territory? Did they care for their young? What did they eat? Did they form herds? Did they communicate amongst themselves by acoustic signals or gestures? Some of these questions have been answered, in a wide variety of ways, in the fantasy story, sometimes in ways that are in keeping with the information available at the time, and more frequently on the basis of the imagination of film directors, writers, and cartoonists. An obvious example of the latter type is the plethora of stories portraying cave- and sea-dwelling dinosaurs.

The world of caverns and caves has traditionally been associated with the realm of prehistory. We should also remember the frequent linking of dragons with caves and caverns. Both factors could influence the description of a troglodyte dinosaur in some way, although there is no evidence to support the hypothesis. In the film *When Dinosaurs Ruled the Earth* (1969), a ceratopsian (probably *Chasmosaurus*) mortally injures a caveman with its sharp supraorbital horns. The animal appears in a fury at the mouth of a cave that the primitive humans were exploring. The shipwrecked space travelers of *Planet of Dinosaurs* (1978) come up against an enormous theropod that has made its lair in a large cave, in which it devours its prey in the same way as do cave bears and lions

In addition, the habitat of the sauropods has been and continues to be one of the most debated aspects of dinosaur ecology. In 1841, the famous British paleontologist Richard Owen described the first known sauropod remains. The Victorian naturalist named this animal *Cetiosaurus*, basing the name on his interpretation of the animal as an enormous reptilian whale that, of course, had aquatic habits. Subsequent discoveries led Owen to alter his point of view partially, although he never abandoned the idea that it was at least partly an aquatic animal. The extraordinary discoveries of U.S. sauropods led Edward Drinker Cope and Othniel Charles Marsh to conclude that these animals had amphibian habits. The most common reconstruction of the camarasaurs and brontosaurs showed them submerged in relatively deep waters, using their long necks as snorkels and periscopes. These ideas broadly prevailed in paleontologic thought until the previously discussed phenomenon of the Dinosaur Renaissance, which advocated essentially terrestrial habits for the sauropods.

Some paleontologists of the first half of the 20th century adopted an extreme position. Sauropods were strictly aquatic forms whose great biomass made it impossible for them to move around on land. Their extremities, away from the gravitational compensation of the water, would literally snap. Cinema, however, has had to discard this radical point of view. The first version of *The Lost World* (1925) portrays a terrestrial sauropod, although one with undoubted swimming abilities in that at the end of the film it leaves London by swimming down the River Thames. The magnificent sauropod animated by Willis O'Brien in *King Kong* (1933) is an animal with aquatic habits that has no problem walking on dry land. In the foggy lagoon of Skull Island, the expedition team, which has set out on a raft, is attacked by a brontosaur. The animal causes the rudimentary craft to capsize when it emerges from deep waters, but it has no difficulty in running actively after the survivors on dry land. The sauropod family in *Baby—Secret of the Lost Legend* (1985) is basically terrestrial. Nevertheless, the parents and the baby are perfectly able to swim. In *One Million Years B.C.* (1966), a brontosaur terrifies Tumak, the primitive human hero, in a desert environment in the Canary Islands. The animal has left enormous impressions of its extremities, even in a dry substrate.

This information indicates that the filmmakers have moved between the model of the "aquatic sauropod that can move on dry land" and the "terrestrial sauropod that can swim." In the 1950s, the Russian paleontologist Ivan Antonovitch Efremov proposed a marine habitat for the sauropods, whose bathymetric distribution would depend on the length of their necks. This heterodox opinion has traditionally been rejected by official dinosaurology. Nevertheless, some dinosaurs may possibly have been more closely related to marine environments than is supposed, as evinced by the discovery of thousands of eggs on a beach from the Upper Cretaceous in Bástus (Lleida, Spain). The episode "Brontia" of the series *A Little Girl to Whom Nothing Ever Happened* by the Russian writer Kiril Bulychev persists with Efremov's idea in representing a live brontosaur, hatched from an egg found on the banks of the River Yenisei, that prefers warm, salty water.

Cinema has accepted without great difficulty the idea of the existence of marine dinosaurs, although some peculiarities should be stressed. Let us first list a few of the main titles: *The Beast from 20,000 Fathoms* (1953), *The Giant Behemoth* (1959), and *Gorgo* (1961). All the dinosaurs portrayed in these films belong to the realm of the paradinosauroids (see chapter 22). Additionally, all these films were directed by the Franco-American director Eugène Lourié (1902–1991), a filmmaker fascinated by the function and effects of giantism in everyday life. The dinosaurs in the first two aforementioned films (Behemoth and the specimen of "*Rhedosaurus*" that devastates New York) are great quadruped beasts with a morphology that is a far cry from that of a marine tetrapod (interestingly, both animals are bathyal or even abyssal). In the course of some underwater sequences, both Behemoth and the "*Rhedosaurus*" swim with agility with the front limbs held close to the body (like a crocodile) and an improbable propulsory use of the back legs. The case of Gorgo is even stranger. A bathyal bipedal tetrapod is one of the strangest combinations that could come into the fevered mind of a biologist.

To describe the habits of a species of animal, as well as investigating where it lives, it is also necessary to know what foodstuffs it consumes, among many other matters. Dinosaurs in the fantasy story can eat anything, as is the case of the "*Rhedosaurus*," which eats a shark, an octopus, and even a New York policeman. The diet of phytophagous dinosaurs is commonly respected by their representatives in the fantasy story, with the specific details that dinosaurology can provide at any given moment—as, for example, in the case of the sauropods. The model of the essentially aquatic sauropod maintains the view that these great animals ate soft aquatic plants. This is the idea that L. Sprague de Camp describes for the sauropods in his previously discussed story, "A Gun for Dinosaur" (1956). The enormous brachyosaurs in *Jurassic Park* (1993) feed on leaves and tree branches, in a manner consistent with the currently prevailing model of essentially terrestrial sauropods. In the novel *Tyrannosaur* (David Drake, 1993) a titanosaurian sauropod adopts a tripod posture (standing upright on its back legs and supported by its tail) in order to be able to feed on the highest branches of a tree, as the U.S. dinosaurologist Robert T. Bakker has suggested.

All paleontologists agree that genera such as *Velociraptor, Cerato-*

saurus, Allosaurus, and *Tyrannosaurus* were carnivores. It seems likely that some types of small- to medium-sized theropod hunted cooperatively in groups, as did the velociraptors in *Jurassic Park.* Some dinosaur experts propose that the big theropods would not have been active hunters but rather carrion feeders. This is not a new point of view. One of the first paleontologists to propose this was Lawrence Lambe, who studied the tyrannosaur *Albertosaurus* during the first years of World War I. Lambe considered that the large ribs of this carnivore and the development of the pubic foot (the distal extension of the pubis) constituted a system by which the animal could rest easily on the midline of its belly. If this were the case, then it was probably so because the animal was slow and sluggish and therefore a scavenger. Lambe developed this proposal at a time when there was a consensus about the hunting habits of the great theropods. Osborn considered *Tyrannosaurus* to be an authentic "killing machine" (especially of *Triceratops*) with a wild and devastating ferocity. These ideas were popularized by the artist Charles R. Knight. In 1906, he produced a picture for the American Museum of Natural History portraying the struggle of a tyrannosaur against a family of *Triceratops.*

The fantasy story has oscillated between these two models, that of the scavenger and that of the hunter. In general terms, the cinema has preferred the latter, probably for purely cinematic reasons: an active hunter will always result in more action and adventure than a boring carrion feeder. In *The Lost World* (1925), an allosaur attacks its prey: brontosaurs and hadrosaurs. And in *King Kong,* the famous struggle between King Kong (1933) and a tyrannosaur is one of the highlights of the film. When the fierce theropod leaps on its prey, Ann Darrow (played by Fay Wray), it is intercepted by Kong. After a memorable fight, with the tyrannosaur's sharp teeth set against the ability and skill of the enormous ape, the great theropod is beaten by Kong. In the film *Fantasia* (1940), a three-fingered tyrannosaur kills a stegosaur to the rhythm of Igor Stravinsky's *The Rite of Spring.* A massive *Ceratosaurus* fights with a *Triceratops* in *One Million Years B.C.* (1966), and in the same film, an allosaur attacks a village of primitive human beings, killing several people. The huge carnivore that is the star of the film *The Valley of Gwangi* (1969) confronts a styracosaur. Many more examples could be mentioned, but all of them illustrate the same point: the large carnivorous dinosaurs of the cinema are terrifying hunters.

A few of today's experts, such as the famous U.S. paleontologist John R. Horner, tend toward believing that the tyrannosaurs were typical scavengers. In the opinion of this well-known Montana paleontologist, an advisor on the film of _Jurassic Park,_ the treatment of the tyrannosaurs in this film is fictitious (according to him, his contribution to the film was merely to ensure that the reconstruction of the dinosaurs was the best that it could possibly be). For Horner, _Tyrannosaurus_ was a heavy animal that could not reach the necessary high speeds for an active hunter. This hypothesis would seem to be reinforced by the fact that in running forms, the tibia is relatively much bigger than the femur. By contrast, these bones are of similar size in the tyrannosaur. The hands would be useless for attacking prey because they are so small and the arms are too short to reach to the dinosaur's own mouth, or even to touch itself around the middle of its chest. An analysis of the tyrannosaur's brain reveals that the optic lobes are small, but the olfactory lobes are quite large. This may mean that it had a strong sense of smell to the detriment of its visual capacity, which, according to Horner, would be typical of a terrestrial scavenger.

In principle, it does not seem necessary for a scavenger to be large in size, as is the case of the tyrannosaur. The reason advanced by Horner is that the size could be of great help in avoiding other hunters or even snatching its prey from them, which were always smaller theropods. Whatever the explanation, some of Horner's arguments are not convincing. The fact that the femur and tibia are the same size does not detract from the fact that its rear limbs were relatively long, which would allow it to move forward in great strides. Furthermore, the construction of the tyrannosaur cranium is closer to what is required for an active hunter than for a scavenger. All things considered, and in the absence of any solid evidence one way or another, it seems reasonable to regard the tyrannosaur as being at the same time both a scavenger and a hunter. Although they may not have been able to actively pursue an off-road vehicle or a rapid ornithomimid, as they do in Spielberg's film _Jurassic Park_ (1993), they could probably have killed some prey, such as the sick and the old. This suggestion is not incompatible with the conception put forward by Horner of the circumstances in which the tyrannosaurs would have hunted. In the famous ornithopod or ceratopsian Bone Beds (accumulations of bones of hundreds or even thousands of individuals in the same level), the discovery of isolated tyrannosaur teeth is a frequent occurrence. There is even one unusual specimen of a hadrosaur bone within which is preserved the sharp point of a tyrannosaur tooth. According to Horner, the tyrannosaurs would have been assiduous companions of the migrations of the enormous herds of plant-eating animals, in which they could have been responsible for an increased number of deaths.

Although cinema, as we have just seen, has tended to consider the tyrannosaurs and other related animals to be hunters, there are a number of stories in science fiction literature that describe the mixed model. For L. Sprague de Camp ("A Gun for Dinosaur," 1956), tyrannosaurs were more devourers of carrion than of live prey. In the novel *Tyrannosaur* (1993), David Drake describes these giant theropods as being more akin to hunters than to scavengers. Their smell is a mixture of snake and carrion. This latter is the result of decomposing meat caught between the teeth. The skull of the famous tyrannosaur specimen "Sue" has a cavity that, according to Horner, could have developed as the result of a mouth infection.

Although interspecific conflicts involving different species (typically a carnivore and a plant eater) are common in fantasy stories, intraspecific conflicts—that is, those involving individuals of the same species—are rare. One of the few cases seen in the cinema appeared in the film *The Lost Continent* (1951), in which two *Triceratops* lock their horns and lift each other, also biting each other with their beaks. Finally, one of them gores the other beneath the neck, killing it. Current ideas concerning the utility of the horns, frills, and other extraordinary protuberances of the ceratopsian dinosaurs favor these types of explanations. Instead of being interpreted as structures for defense against predators, they are considered to be resources involved in the competition for obtaining mates, or as structures with a display function (like a peacock's tail feathers), or as organs for terrifying possible competitors (like a wolf's teeth). However, unlike the mortal combat engaged in by the two *Triceratops* of *The Lost Continent*, it is possible that these conflicts were ritualized, a common strategy in Nature to avoid the deaths of the adversaries.

The ceratopsians and other dinosaurs would probably have had sophisticated rules of behavior that are consistent with the fact that they were gregarious animals that associated in herds. It is known from the information from the Bone Beds that many dinosaurs, such as the ceratopsians and the hadrosaurs, were gregarious. In general, this habit has not been widely portrayed in the cinema, probably because of the difficulties that it presents for traditional special effects techniques. On the other hand, however, comics and science fiction tales have reflected this phenomenon. In *Doctor Who and the Dinosaur Invasion* (Malcolm Hulke, 1978), the novelization of the *Doctor Who* program "Invasion of the Dinosaurs" (1974), the *Triceratops* move around North America in large herds. The book also supplies further fascinating information: the beasts could charge at speeds of more than 48 km/h and impale

The Lost Continent (Lippert, 1951). This movie has the distinction of being less famous than its re-dubbed spoof (*Mystery Science Theater 3000,* Best Brains Productions).

their prey with their horns! The U.S. author Harry Turtledove's story "Hatchling Season" (1985) faithfully picks up what was believed about the behavior of hadrosaurs in the mid-1980s. A young paleontologist goes to the Cretaceous in Montana. She has two weeks to finish a doctoral thesis on the behavior of these duck-billed dinosaurs. The hadrosaurs are, naturally, gregarious, and the main interest of the story centers on the field observations that the dinosaurologist makes about the reproductive biology of the hadrosaurs (see chapter 26).

A tick bites the paleontologist of "Hatchling Season." There is no evidence in the fossil record that dinosaur ectoparasites, such as ticks or other arthropods, existed. However, the idea is perfectly plausible and has been included in various stories of the fantastic discourse. In "A Gun for Dinosaur," a thumb-sized tick tries to parasitize one of the members of the time safari. The story "Poor Little Warrior" (1958) by Brian W. Aldiss describes the death of a sauropod caused by a hunter from the 22nd century. The time traveler manages to kill the enormous dinosaur but dies at the same time as a result of the giant lice that abandon their host upon its death. The dinosaur ectoparasites in Aldiss's story are certainly colossal—a highly unlikely idea shared by the film *Godzilla 1985* (1985), in which an ectoparasite of the popular Japanese monster, looking very like a trilobite, measures about a meter in length.

In this chapter, I have shown how various aspects of dinosaur biology—for example, habitat, diet, and behavior—are represented in the fantasy story. Other aspects that have to do with their reproductive biology, such as egg-laying behavior, egg size, and possible parental care of the clutch and offspring, deserve their own chapter.

26

Eggs and Young

The fossil record has limitations that are increasingly better understood. The work of a good paleontologist is to assess these limitations as perfectly as possible and to extract consistent and objective information from the fossil evidence. These premises are especially evident when we try to establish what the reproductive habits of animals from our distant past—in this case, dinosaurs—would have been. Initially, it seems logical to think that the dinosaurs had some kind of specific egg-laying and parental care behavior. We should keep in mind that the nonavian dinosaurs are situated phylogenetically between birds and crocodiles. As we know, both groups have patterns of behavior that are often sophisticated with respect to reproductive biology. Therefore, we may suppose that the nonavian dinosaurs also had them.

The discoveries made by staff members of the American Museum in Mongolia during the 1920s confirmed that the dinosaurs could have had complex patterns of nesting behavior (e.g., nests made of concentric strings of eggs) but told us nothing of possible parental care. This knowledge comes initially from the discoveries of John R. Horner in Montana at the end of 1970s. This period marked a genuine revolution in the knowledge of dinosaur reproductive behavior: for the first time, there was reliable evidence that at least some species cared for their offspring. Strangely, cinema had adopted this view years before there was any evidence from the fossil record that supported this notion.

One of the first films to incorporate dinosaur young was *The Lost World* (1925). The members of the expedition watch a mother *Triceratops* actively defending its young against a large carnivore. The female finally manages to drive off the predator and licks its young lovingly. The cinema technicians responsible for the dinosaurs in *The Lost*

World (Willis O'Brien and Marcel Delgado) teamed up again in 1930 to make the film *Creation*, which was never finished. In this film, a callous hunter kills a young *Triceratops*. The evil character is hunted down and killed by the furious mother. In 1956, O'Brien built the models of the dinosaurs for the film *The Animal World* (1955), based on Charles R. Knight's illustrations and under the supervision of Charles L. Camp. The visual effects technician Ray Harryhausen took charge of animating the dinosaurs. *The Animal World* is a documentary about present-day fauna that includes 15 minutes of footage about prehistoric life during the Jurassic and Cretaceous. A female *Apatosaurus* lays its eggs and is then frightened by a large theropod (note that other cinematic dinosaurs, such as the ceratopsians, would probably have attacked the predator). Finally, the sauropod's eggs begin to hatch.

One of the most interesting dinosaur mothers in the history of cinema appears in *Gorgo* (1961). An underwater explosion liberates a giant creature off the Irish coast. Gorgo, as we have already seen, is a huge biped, with a well-developed dorsal exoskeleton and projections from its head similar to the fins of a fish. The monster is taken to London to be exhibited in a circus. Another gigantic monster, the 400-m-tall mother, follows it to the British capital, which is virtually demolished by the fury of the mother rescuing its 105-m-tall offspring. Famous London monuments such as the Tower Bridge and Big Ben are reduced to ruins. In the end, the mother rescues its baby, and they return to the depths of the oceans, leaving man to reconsider his proud belief that he is the lord of creation.

The Japanese film *Gappa, the Triphibian Monster* (1967) has a similar structure. A volcanic eruption incubates an enormous egg on a South Pacific island. The hatchling (a *Gappa*) is captured by scientists and journalists, who take it to a laboratory in Tokyo. Dad and Mom *Gappa* set off for Japan to rescue their baby. The *Gappas* are enormous winged bipeds that can fly and dive. They have toothed beaks, and the scientists discover that they are related to birds. The *Gappa* parents destroy a number of Japanese tourist attractions before at last finding their offspring, which has been set free in Tokyo airport. The young monster hugs its parents, who then teach it to fly. The family then take off in the direction of their South Pacific home.

In *When Dinosaurs Ruled the Earth* (1969), a cavewoman (the actress Victoria Vetri in the role of Sanna) takes refuge in a fragment of the shell of an enormous egg during a nighttime storm. The next morning, a giant female dinosaur (a paradinosauroid; see chapter 22) finds the woman and mistakes her for one of its own young. The surprised dinosaur offers her a deer for her breakfast. Sanna lives with her step-brother, a baby dinosaur whom she trains with a flute. The baby is portrayed with a great sense of humor, based on the behavior of a puppy.

The plot of the film *Baby—Secret of the Lost Legend* (1985) has already been described. The behavior of the young brontosaur is characterized by mammalian and even anthropomorphic traits. The film is about an orphan, and the story stresses the association of the character with the so-called Bambi Syndrome. The young dinosaur cries in front of the dead body of its father, an unoriginal device aimed at heightening the emotions of the spectator. The degree of anthropomorphization of animal characters has traditionally been accentuated in animated cartoons. The film *The Land before Time* (1988) is a further example of this trend. The story again makes use of the Bambi Syndrome as a simple semiologic device and gives each of the dinosaur young its own human nature. Only the evil character, an adult tyrannosaur, lacks this human condition (it is apparently unable to speak).

The opposite of these mystifications about the behavior of dinosaur parents and offspring is the literary science fiction story, which has tended to adopt the scientific information available at the time, as in the case of Harry Turtledove's story "Hatchling Season" (1985). The hadrosaurs formed large nesting colonies to which they kept returning and fed their young in a similar way as do seabirds.

In addition to caring for their young, dinosaurs in the fantasy story can also defend their eggs. This is the case with an adult *Triceratops* that actively defends its nest from the attack of a ceratosaur in *The Land That Time Forgot*. This ceratopsian's nest has four or five big eggs. Large-sized or even huge eggs are one of the typical features of cinema dinosaurs. In *Caveman* (1981), an enormous pterosaur egg is cooked in a type of small volcanic thermal spring. The primitive humans cut the boiled egg into slices and take them back to their settlement. A number of the aforementioned Japanese *Kaiju Eiga* films feature gigantic eggs. One of the best is *Rodan* (1956), in which a pair of enormous pterosaurs appear in a coal mine after the incubation of some enormous eggs. A paleontologist studies a fragment of the shell, and his main conclusions are that the egg is more than 20 million years old, that its microscopic structure is typical of reptiles, and that its original volume was some 2,830 m³.

As we saw in chapter 11, an egg is sometimes the cause of an encounter between humans and dinosaurs. This happens in two particular films, the Spanish film *The Sound of Horror* (1965) and *Twenty Million Miles to Earth* (1957). The first has a confusing beginning, probably the product of the lack of discrimination between archeology and paleontology. A group of archeologists find a dinosaur egg in some Greek ruins. The animal, which is invisible, is finally destroyed by fire.

The most famous exodinosaur in the cinema is Ymir (the Nordic name for the mythological father of the giants) in the film *Twenty Million Miles to Earth*. It is a bipedal creature with the appearance of a theropod and has tridactyl hands and feet. The exodinosaur comes from Venus and is brought to Earth on a rocket. It hatches out of an elongated egg composed of a gelatinous substance and proceeds to grow rapidly, quadrupling in size to a height of 1.2 m in 24 hours.

We can conclude that the behavior of the parents and the offspring of many of the dinosaurs in the cinema has been characterized by its mammalian and even anthropomorphic traits from cinema's beginnings. Certain titles, such as *Gorgo* and *Gappa*, have a standard structure: (1) the young dinosaur is captured; (2) the parents set out to rescue it and attack the humans; and (3) the young dinosaur is rescued and the family of monsters returns home. *Gorgo* and *Gappa* symbolize Nature's response to human aggression. In this case, the moral is quite clear: Nature defeats the secular human tyranny over the living world. *Gorgo*, *Gappa*, and also *Baby* go beyond the simple portrayal of mammal-like parental care and defend the family as a functional unit in the most anthropomorphic sense. Finally, the majority of dinosaur films that include eggs portray them as being huge. This extraordinary idea may be aimed at heightening the audience's surprise, but may be based on the two popular beliefs that prehistoric animals were gigantic, powerful creatures and that huge animals should lay really large eggs.

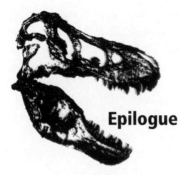

Epilogue

Dinosaurology at the beginning of the 21st century is an active and dynamic business, whose continued significance in the new millennium is guaranteed. The U.S. paleontologist Peter Dodson has calculated that the total number of dinosaur genera that have existed is somewhere between 900 and 1,200. If this claim is correct, then we currently know only 25% to 30% of the total generic diversity of the dinosaurs. In any case, the future of research seems to be assured. But why should we study dinosaurs? A simple answer is that they form an important part of our planet's past. The study of dinosaurs can provide important information about specific phenomena that regulate the evolutionary history of living organisms, especially those relating to communities of large terrestrial animals, whose only contemporary point of reference is basically the mammals.

Furthermore, paleontology, just as any other branch of scientific knowledge, has an indubitably sociocultural responsibility. Dinosaurology is one of the areas of paleontology that has been most deeply committed to this. Society's immense curiosity about dinosaurs has to be duly satisfied. At the same time, this social pressure ensures that there is a daily increase in the amount of dinosaur research. This complex feedback process is one of the specific characteristics that has ensured the establishment and popularity of dinosaurs in society.

As an illustration of the complex interaction between dinosaurology and society, I will provide a few examples. The paleontologist Bruce McFadden has indicated that the courses on dinosaurs in various U.S. university geology departments has allowed them to guarantee their funding. Toy manufacturing companies have hired dinosaurologists to contribute to the realistic design of plastic dinosaurs. Finally, some

Dinomania used as a means of cultural attraction and for exploiting possibilities for tourism: The life-size model of Spanish sauropod *Aragosaurus* on the day of its arrival in the village of Galve (Teruel, Spain) at the end of 1993.

Road sign of the Government of La Rioja (Spain), used to attract tourists to the region's famous beds of dinosaur tracks.

groups have become convinced that they should carry out their activities concerning the local dinosaur record in a manner that balances two interests: on one hand, the responsibilities for protecting and conserving the fossil heritage, and on the other, its use as a basis for the development of an educational and sustainable tourist industry.

In recent years, the media has frequently reiterated that dinomania is due to the *Jurassic Park* phenomenon. In truth, it started in the 19th century and has since proceeded in fits and starts. Dinomania cannot be evaluated as a passing sociocultural fad, as has happened with, for example, angels, or the flared trousers of the 1960s. Dinosaurs are firmly rooted in popular culture; they constitute one of the clearest areas of interaction between scientific information and the public's thirst for knowledge—in this instance, knowledge concerning life in the past. The thrill and fascination humanity feels toward dinosaurs will persist as long as we continue to exist.

INDEX

Page numbers in italics indicate illustrations.

José Luis Sanz is Professor of Paleontology at the Universidad Autónoma de Madrid (Spain). Author of six books and more than 130 technical articles, Sanz conducts research on dinosaurs and other Mesozoic reptiles in Western Europe. He has named four new genera of dinosaurs and three primitive birds and now focuses his research on the early evolutionary history of birds, their phylogenic relationships, and the origin of flight.